SpringerBriefs in Applied Sciences and Technology

Computational Intelligence

Series editor

Janusz Kacprzyk, Polish Academy of Sciences, Systems Research Institute, Warsaw, Poland

The series "Studies in Computational Intelligence" (SCI) publishes new developments and advances in the various areas of computational intelligence—quickly and with a high quality. The intent is to cover the theory, applications, and design methods of computational intelligence, as embedded in the fields of engineering, computer science, physics and life sciences, as well as the methodologies behind them. The series contains monographs, lecture notes and edited volumes in computational intelligence spanning the areas of neural networks, connectionist systems, genetic algorithms, evolutionary computation, artificial intelligence, cellular automata, self-organizing systems, soft computing, fuzzy systems, and hybrid intelligent systems. Of particular value to both the contributors and the readership are the short publication timeframe and the world-wide distribution, which enable both wide and rapid dissemination of research output.

More information about this series at http://www.springer.com/series/10618

Soumya Sen · Anjan Dutta · Nilanjan Dey

Audio Processing and Speech Recognition

Concepts, Techniques and Research Overviews

Springer

Soumya Sen
A.K. Choudhury School of Information
Technology
University of Calcutta
Kolkata, West Bengal, India

Anjan Dutta
Department of Information Technology
Techno India College of Technology
Kolkata, West Bengal, India

Nilanjan Dey
Department of Information Technology
Techno India College of Technology
Kolkata, West Bengal, India

ISSN 2191-530X   ISSN 2191-5318   (electronic)
SpringerBriefs in Applied Sciences and Technology
ISSN 2625-3704   ISSN 2625-3712   (electronic)
SpringerBriefs in Computational Intelligence
ISBN 978-981-13-6097-8   ISBN 978-981-13-6098-5   (eBook)
https://doi.org/10.1007/978-981-13-6098-5

Library of Congress Control Number: 2018967442

This Springer imprint is published by the registered company Springer Nature Singapore Pte Ltd.
The registered company address is: 152 Beach Road, #21-01/04 Gateway East, Singapore 189721,
Singapore

The Human Voice is the most perfect instrument of all

—Arvo Pärt

Preface

In recent years, a rapid advancement is seen in the audio communication methodology. The objective of communication methodology is to provide a continuous seamless, easy, reliable, and high-quality communication between people and machines. Human-to-machine interactions are done using different mediums such as text, graphics, touch screen, mouse, and speech; however, speech is the most intuitive and natural communication method.

Objective of the Book

This book has been designed to give the basic idea of audio processing and gradually advances toward the methodologies used in audio processing and speech recognition. Furthermore, it gives the ideas on the ongoing research approaches related to audio processing. The focus of this book is to cover a wide range of audience including the advanced readers like researchers, postgraduate students, and also the beginners to provide a fair idea about different aspects of audio/speech processing and speech recognition. The purpose of this book is to highlight and discuss the published research in the areas of audio processing and speech recognition and to stimulate further research interests for innovations.

Organization of the Book

This book consists of four chapters, detailing the concepts and techniques of audio processing and speech recognition. In Chap.1, audio indexing and its importance in audio searching are discussed. In Chap. 2, a brief idea of speech processing and electrical model of human speech production system is given. The architecture of a typical automatic speech recognition (ASR) system is also detailed here. Feature extraction is an important phase of any ASR system. In Chap. 3, a vivid idea about the basic audio features and feature extraction techniques is discussed. After the feature extraction, speech/audio classifications are done based on the extracted features. In Chap. 4, some state-of-the-art classification techniques and related research on speech classification are discussed.

Chapter 1: Audio Indexing

In this era of information technology, there is an exponential growth of information content from various sources. A huge amount of digital audio, video, and images are being generated in a stupendous rate. In order to effectively use this large amount of multimedia data, there should be some efficient search methodology so that the desired information could be easily obtained for the decision-making process. The challenges of information retrieval from a large audio file and the importance of audio indexing are discussed here. Audio indexing is of two types: text-based indexing or large vocabulary continuous speech recognition (LVCSR) and phoneme-based indexing. In LVCSR system, the entire audio is first transformed into textual form and after that the information retrieval algorithms are applied to index the textual data. Phonemes are the basic building block of speech. In this approach, a phonetic-based index is created from an audio file. After that, the users' search term is converted into possible phoneme string with the help of a predefined phonetic dictionary. The search term then is looked in the index. The detailed discussion of LVCSR and phoneme-based indexing and their pros and cons are detailed here. This chapter ends with by providing a comparative study between LVCSR system and phoneme-based indexing.

Chapter 2: Speech Processing and Recognition System

Speech processing is a special case of digital signal processing (DSP) which is applied to process and analyze speech signals. Some of the typical applications of speech processing are speech recognition, speech coding, speaker authentication, speech enhancement, that is, detection and removal of noise, speech synthesis, that is, text-to-speech conversion. Knowledge of human speech production system is the prerequisite of speech recognition task. A brief discussion of the human speech production system and corresponding method of developing an electrical speech production system is discussed in this chapter. A brief history of the evaluation of ASR system is discussed so that readers could have a fair idea about the root of ASR and subsequent developments. A typical ASR system consists of acoustic analysis, acoustic model, pronunciation model, and language model. The input sound waveform is transformed into some discrete feature vectors in acoustic analysis stage. The relationship between the speech signal and phonemes is represented by an acoustic model. Here, some training algorithm is applied to the speech database to create a statistical representation of each phoneme which is also known as hidden Markov model (HMM). In recent years, there is a revolution in speech recognition with the advent of neural network and deep learning system. It is observed that neural network with the combination of HMM has increased the accuracy of speech recognition significantly. The final output of the acoustic model is a set of phonemes with their probability measures. The mapping of the phonemes to the specific words is done by pronunciation model. Words are organized to form a meaningful sentence in the language model. The detailed description of each component of the ASR system, their utility, and interdependence is discussed in this chapter.

Chapter 3: Feature Extraction

The prerequisite to classify an audio or speech signal is the feature extraction. Different audio features and corresponding feature extraction techniques are discussed in this chapter. This chapter begins from explaining very basic concepts like framing and gradually digs deep into the various audio features. Different sound sources could be distinguished by pitch. Different sounds having same pitch and tone could be identified by timbral features. Some of the important timbral features are zero crossings, spectral centroid, spectral roll-off, and spectral flux. Rhythmic features are defined by the regularity of the audio signal. Two important similarity measures—inharmonicity and autocorrelation—are discussed here. Deviation of an audio signal from a purely harmonic one could be measured by the feature inharmonicity. Autocorrelation is an important similarity measure between two audio signals. There is a need to standardize the features for maintaining uniformity in the feature extraction techniques. This chapter gives a vivid description of MPEG-7 standard features. Various speech extraction techniques are discussed in detail here. Linear predictive coding (LPC) is used to estimate the current speech sample by analyzing the previous speech sample. The speech signal is represented by some parametric form through Mel-frequency cepstral coefficient (MFCC). MFCC uses mel scale. The frequency sensitiveness of the human auditory system could be replicated by mel scale. Perceptual linear prediction (PLP) considers the human perception of speech. The above feature extraction techniques are applicable only for analyzing the stationary signals, but in real time, one has to work with non-stationary speech signals. After discussing the standard techniques like LPC, MFCC, and PLP, limitations of Fourier and short-time Fourier transform (STFT) are discussed for analyzing nonstationary audio signal. Wavelet and discrete wavelet transform could be used in order to analyze time-varying audio signal. The evolution of the wavelet transforms from Fourier transform through STFT, and a brief idea of it is given at the end of this chapter.

Chapter 4: Audio Classification

Classification is the process of categorizing items in a dataset to individual classes based on its characteristics or features. The classification phase in ASR is one of the most highlighted and attractive research domains. Some popular classification techniques and their application in automatic speech recognition are illustrated in this chapter. The traditional classification techniques and related research work in speech classification are discussed in detail. K-nearest neighbor (k-NN) is the age-old classification technique which still has relevance in the modern era of classification. k-NN is suitable for the systems where there is a little prior knowledge of the distribution of data. After discussing some limitation of k-NN, Naïve Bayes classification is discussed. Naïve Bayes is a much faster method than k-NN. Conditional independence among the features is considered in Naïve Bayes, and a maximum likelihood hypothesis is used. Decision tree is another state-of-the-art nonparametric supervised classification technique. Based on some preconditions, an attribute is selected to split the dataset and ultimately a decision tree is generated. Decision rules could be formed by analyzing the decision tree. One of the popular supervised machine learning algorithms is support vector machine (SVM) which could be effectively used for the classification of a multidimensional dataset. A boundary is fixed to a collection of data points which are all alike or belong to the same class. A new data point is compared with the boundary to know its class label. This boundary is also called a hyperplane. All the above-mentioned traditional classification techniques have been discussed in this book, and at the end of each section, some of the pioneering-related research works are discussed. With the advent of neural network, a revolution in ASR system is observed. Neural network model could be used as an alternative to traditional classifiers. It is seen that if neural network is used in combination with the HMM model, then there is a significant improvement in the accuracy of the ASR system. Some of the pioneering researches on using neural network classifier in speech recognition are briefed in this chapter. In the last decade, a deep neural network has been evolved as an interesting research domain. Several researches have been done on the application of a deep neural network in speech recognition; some of these are discussed at the end of this chapter.

We are thankful to our families for their continuous support for completing this book. We would like to express our gratitude to Mr. Prasenjit Banerjee for helping us to create some high-resolution figures and diagrams.

Kolkata, India   Soumya Sen
                                                                             Assistant Professor
                                                                                  Anjan Dutta
                                                                             Assistant Professor
                                                                                 Nilanjan Dey
                                                                             Assistant Professor

Contents

About the Authors

Soumya Sen is an Assistant Professor at A. K. Choudhury School of Information Technology, University of Calcutta. He received his Ph.D. (Tech) degree from the Department of Computer Science & Engineering, at the same university, in 2016. Before joining A. K. Choudhury School of Information Technology, he worked at IBM India Pvt. Ltd and RS Software. His industrial expertise includes ERP & data warehousing. Currently his research interests are data warehousing & OLAP tools, data mining, big data, service engineering, distributed databases, and machine learning. He has published 1 book, 70 research papers in peer-reviewed journals and international conferences and registered 3 patents in USA, Japan & South Korea. Dr. Sen is a PC member and reviewer for numerous International conferences.

Anjan Dutta was born in Kolkata, India, in 1986. He received his B.Tech. degree in Information Technology from the West Bengal University of Technology in 2008 and M.Tech. in Information Technology in 2011 from Calcutta University.

He served in IXIA Technologies Ltd and TATA Consultancy Services Ltd (TCSL) over 6 years of period. Initially he worked as a protocol developer in IXIA Technologies Ltd and worked on 3gpp wireless protocols. Thereafter he worked as an IT Analyst in TATA Consultancy Services Ltd (TCSL) from July 2011 to

July 2017. He is now employed as an Assistant Professor in the Department of Information Technology, Techno India College of Technology, India. He is an active researcher in the field of Big Data, Data Mining, Audio processing and Audio classification etc.

Nilanjan Dey was born in Kolkata, India, in 1984. He received his B.Tech. degree in Information Technology from the West Bengal University of Technology in 2005, M.Tech. in Information Technology in 2011 from the same University and Ph.D. in digital image processing in 2015 from Jadavpur University, India.

In 2011, he was appointed as an Assistant Professor in the Department of Information Technology at JIS College of Engineering, Kalyani, India followed by Bengal College of Engineering College, Durgapur, India in 2014. He is now employed as an Assistant Professor in the Department of Information Technology, Techno India College of Technology, India. He is a visiting fellow of the University of Reading, UK. His research topic is signal processing, machine learning and information security.

Dr. Dey is an Associate Editor of IEEE ACCESS and is currently the Editor-in-Chief of the International Journal of Ambient Computing and Intelligence, Series Co-editor of Advances in Ubiquitous Sensing Applications for Healthcare (AUSAH), Elsevier and Springer Tracts in Nature-Inspired Computing (STNIC).

Chapter 1
Audio Indexing

1.1 Introduction

Audio is available from various sources like recordings of meetings, newscast, telephonic conversations, etc. In this era of information technology, with the technological progress, more and more digital audio, video, and images are being captured and stored day by day. The amount of audio data is increasing exponentially on the web and other information storehouses. In order to efficiently use this huge multimedia data, there should be an effective search technique. In earlier days it was difficult to interpret, recognize and analyze the digitized audio using computers. The audio piece was converted into textual representation, and manual analysis was done on it. However, gradually with the advent of larger storage capacities, superior microprocessors, and state-of-the-art speech recognition algorithms, it is possible to extract and index audio content using audio mining. Audio indexing [1] helps to represent the audio documents in a better way by finding a quality descriptor that could be used as indexes to assist in audio search and archiving. In Sect. 1.2, audio indexing and classical information retrieval problem is discussed. Here, why an effective audio search technique is required is discussed. There are two approaches of audio indexing: text-based indexing or large vocabulary continuous speech recognition (LVCSR) and phoneme-based audio indexing. In the former case, the entire audio is first converted into text; after that, the word is searched in a predefined dictionary. In Sect. 1.3, LVCSR approach is discussed in detail. LVCSR approach suffers from mainly two problems: recognition error and limited vocabulary and out-of-vocabulary problem. To overcome that, phoneme-based approach is introduced where the search term is broken into individual phonemes. This is discussed in Sect. 1.4. In Sect. 1.5, a comparative study of LVCSR and phoneme-based approach is discussed.

S. Sen et al., *Audio Processing and Speech Recognition*, SpringerBriefs in
Computational Intelligence, https://doi.org/10.1007/978-981-13-6098-5_1

1.2 Audio Indexing and Classic Information Retrieval Problem

The classic information retrieval problem [2] deals with locating the target text documents, consisting of the search key. If a document has a higher number of query terms, then it is considered as more relevant. The number of common words was counted, and the measurement of text similarity was done. This usual term matching approach is not applicable for audio because of the lack of identifiable words in audio files. There is also an issue, regarding the linearity of audio files. The audio file should be processed entirely from start to finish, ensuring that any important information is not missed.

A human can distinguish between various types of audio effectively. For any audio item, we can have a fair idea about the type of the audio (i.e., whether it is noise or music or human voice) or what is the speed of the audio and mood of the audio, i.e., whether it is in a relaxing tone or not. However, to the perspective of the computer, it is only a sequence of binary data stream. Now, there might be a situation where it is required to know when a particular person has spoken in the entire duration of the audio. Earlier the most common method of searching the audio was by the name of the audio file. But for the above mentioned query, an effective searching technique is needed.

Content-based audio retrieval techniques are needed to solve the above problem. A sample-to-sample comparison of the search query and the audio piece was done, but it was not that effective because the different audio signals have different sampling rates. Because of this, recent research on content-based retrieval technique focuses on extracted audio features [3].

Audio indexing can be defined as an analytical method [4] by which an entire audio file can be processed, and as a result, an index of individual words present in the audio file and their locations are created. This index should be readily searchable. In Fig. 1.1, a high-level view of role of audio indexing in audio search process is shown.

There are two approaches in audio indexing [5]:

 (i) Text-based indexing or large vocabulary continuous speech recognition (LVCSR) and
(ii) Phoneme-based indexing.

1.3 Large Vocabulary Continuous Speech Recognition (LVCSR)

In text-based indexing or large vocabulary continuous speech recognition (LVCSR) system, the entire audio is first converted into text [6]. A dictionary having a huge number of entries is maintained. A word could be searched and identified from this

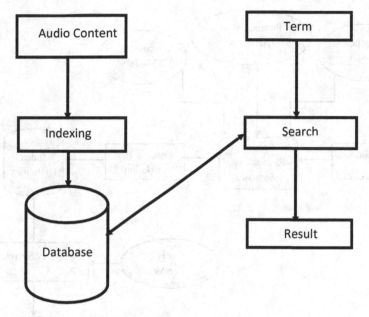

Fig. 1.1 Audio indexing at a high-level view

dictionary. Thereafter, information retrieval algorithms [7] are used to index result-ing textual representation. The search query in word terms could be then used to find the required portion of the audio document. If the text or word could not be found in the dictionary, the LVCSR search engine will do a similarity matching to map the word with a nearly matching entry which resembles it. LVCSR uses a language understanding model to generate a confidence level for the search outcome. If the confidence level of the findings is less than 100%, then it gives options for other sim-ilar word matches. So this system can enhance the accuracy level by storing similar sounded words, although some wrong results also might be generated. Figure 1.2 represents an LVCSR system.

An example of the real-time application of an LVCSR system could be analyzing of customer calls in a call center. It is beneficial for the business to know common customer interest or customer trends. Figure 1.3 explains the above.

The two major disadvantages of LVCSR system are: recognition errors and vocab-ulary limitations and out-of-vocabulary problem [8].

1.3.1 Recognition Errors and Vocabulary Limitations

Audio is inherently random in nature. Different types of speakers and signal states prevail in a particular audio content. The challenge of speech recognition also increases due to external noise.

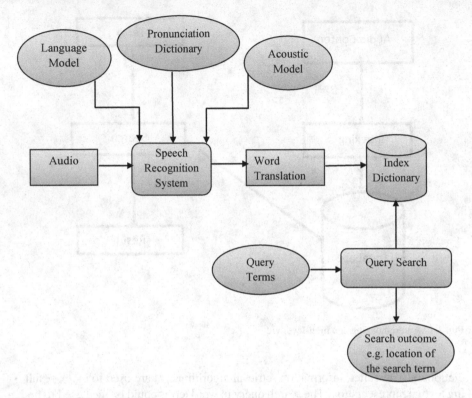

Fig. 1.2 LVCSR system

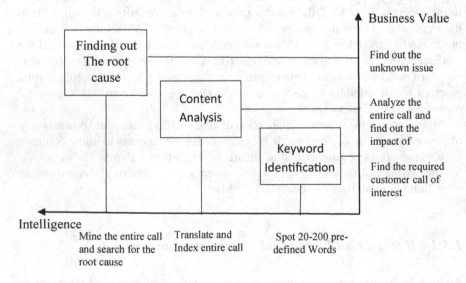

Fig. 1.3 Customer call analysis system

For noisy or conversational communication, the typical recognition error rates may range from 5 to 50%.

Recognition errors could be minimized by several means:

- The inherent nature of the query terms is of having longer words which itself tends to be recognized correctly.
- Training the recognizer by simulating the real-time audio environment. In this way, recognizer can give a fair accuracy rate in a noisy environment also.
- Finally, some words, e.g., names and places, are repeated in an audio program several times, and hence, the probability of recognizing at least one word is high.

Hewlett Packard's SpeechBot system [9–11] demonstrated that a fair search performance could be achieved in a highly erroneous environment also. It is shown that if the error threshold is 25%, then the search engine's performance is similar to those which are acting on the textual transcription of the audio content. But beyond this threshold, the search performance is declining gradually.

1.3.2 The Out-of-Vocabulary Problem

Out-of-vocabulary (OOV) words are another serious problem in LVCSR system or word-based audio indexing. New words are generated continuously with the changing time and the dictionary and language model could not effectively cope up with this. Since they are of insufficient size, they are not suitable for the dynamically changing environment. The nature of LVCSR is static because it relies on a predefined dictionary for speech recognition. Now if a word is not in the dictionary, there is no way to identify or predict the same. One solution to this problem is to expand the dictionary to cover the new words. In this approach, there is a chance of adding acoustically ambiguous words. One approach to solve the above problem is to continuously update the language model and the static dictionary with the current topics if there is a scope to get a suitable source of the textual data. In this way, the rate of OOV in audio content recognition could be effectively reduced. This approach might be useful by making the dictionary more dynamic but the OOV problem remains in audio search queries. In a study related to HP's (earlier Compaq) SpeechBot system, it is seen that OOV rate in user queries are above 10% [11]. Due to this problem, there was a need to find the alternatives for the text-based recognition approach [12].

1.3.3 Pros and Cons of LVCSR Speech Analytics

Keeping in mind of the above two points, the pros and cons of LVCSR system could be summarized. In LVCSR, statistical methods are used to predict the likelihood of different word sequences; hence the accuracy level is much higher than the single word lookup of a phonetic search. So here if a word is found, then the probability of it is actually spoken is very high.

But here the searched words or phrases should be there in the dictionary in order to be found by the recognition system. This is a major disadvantage of LVCSR. In rare cases, if there is no exact map with the word and the dictionary content, then sometimes it could be found by combining more than one words (e.g., call Paul for Calpol).

Since the vocabulary is very large, the initial processing of the audio samples might take a considerable, amount of time. But the search is very fast and accurate because simple test to text matching is checked.

In a situation where a call center manager is trying to find how many times "ABC" company's name is spoken by the customers, both search speed and accuracy need to be taken care of. Faster search speed means, more customer calls could be analyzed in a shorter amount of time, with less human resources cost. If the search is not accurate, then it might affect critical business decisions. So in this context LVCSR system could be applied.

A much higher accuracy is observed in the LVCSR systems since there is a high probability of finding the words that were actually uttered. LVCSR system suffers from lower recall rate due to unusual words or recognition errors. To reduce this issue a key term is found and the transcription of words around the key term is provided. All these transcriptions are visually skimmed by the end user and the relevance of these occurrences could be determined. Automated analysis of the frequency of different words and phrases could be done due to the availability of the actual transcript. New hidden trends and metrics could be revealed that might not be found by the system usually.

1.4   Phonetic Search

The most granular unit of speech is the phonemes. Different kinds of utterances (such as long "b" or long "a" sound) could be distinguished by phonemes. Each and every word is a set of phonemes [13]. Phoneme-based indexing works with sounds instead of converting speech to text. In this approach, first, the sounds in an audio piece are analyzed to create a phonetic-based index. After that, a dictionary containing several dozen phonemes is used to convert users' search term to the possible phoneme string. Thereafter the search term is looked in the index. The phoneme-based search system is more complex than the text-based search.

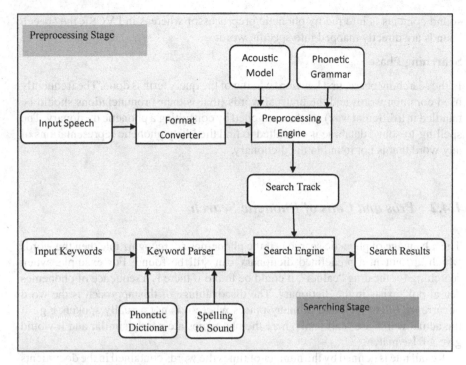

Fig. 1.4 Two-phase phoneme-based search

1.4.1 Phases of Phonetic Search

There are two phases [8] in this approach- preprocessing and searching. The first phase involves in preprocessing the search query to produce a phonetic search track from the input speech. In the second phase, phonetic search track is searched to find the query term(s). Figure 1.4 depicts these two phases.

Preprocessing stage

The format conversion module is used to convert the input speech into a standard format or representation. Thereafter the input speech is scanned by the preprocessor by using an acoustic model and phonetic grammar. The output of this stage is the phonetic search track. The environment in which the words are spoken and recorded by a transducer is known as an acoustic channel. Natural language is used by human beings to express the input speech. The properties of both of an acoustic channel and a natural language are represented by an acoustic model.

A phonetic grammar is used to identify the probable end points of the words in the input speech. A phonetic grammar is dependent on the spoken natural language.

The highly compressed representation of the phonetic content of the input speech is the phonetic search track. The probability of occurrences of the potential phonetic sound contents is inferred by phonetic preprocessor whereas in LVCSR the speech sounds are directly mapped into specific words.

Searching Phase

In the searching phase, first keyword parsing of the query term is done. The frequently used common words and the unusual words (that is their pronunciations should be handled in a different way) are then searched by consulting a phonetic dictionary. The spelling-to-sound database is consulted to find the likely phonetic representations of any word that is not found in the dictionary.

1.4.2   Pros and Cons of Phonetic Search

The advantage of phonetic search is if the phonemes are recognizable then the words which are not in a predefined dictionary can still be found. For example, when searching for the drug "calpol", it could be found if there is a sequence of phonemes "ca al pol" exists in the dictionary. The disadvantage of this approach is the word occurrence might be found in many places where it is not actually spoken, e.g., if the actual words are "call paul", here the phonemes are nearly similar and it would give a false match.

Recall rate is defined by the number of times the words contained in the documents are found by using a search method. Generally, a higher recall rate is observed in the phonetic approach. As mentioned above, this higher recall rate could be influenced by the false positive matches and the same should be filtered out manually from the result set.

Phonetic system processing is faster because the vocabulary size is itself very small because it contains phonemes and there are a few unique phonemes in every language. But since the phonemes cannot be effectively indexed like a whole word, the overall search process is slow. The phonetic system also takes a much larger memory to store the phonemes of spoken words. For a big project, it might be an issue.

The possible sequences of sound and their corresponding frequency are considered in phonetic approaches (for example, "qtldr" which is a group of consonants never appear in English). The higher level knowledge of the language is not considered by the phonetic approach. The phoneme-based model could not filter out the most likely used phrases correctly, resulting in more manual work in later stages.

1.5   Comparison Between LVCSR and Phonetic Search

From the above discussion, it is observed that both LVCSR and phonetic search have some pros and cons. Here, a comparative study is done between them.

- **Accuracy and search time**: As mentioned earlier, phonemes could not be indexed effectively like whole word; therefore, the process becomes slow. A higher recall rate and lower precision are observed in phonetic approaches, whereas LVCSR has lower recall rate and high precision. In phonetic approach, the probability of getting a false match is much higher than the LVCSR system.

- **Vocabulary**: The LVCSR system can only recognize words which are present in the vocabulary. So many words could not be searched due to this. Phonetic preprocessing maintains an open vocabulary and largely unconcerned about such kinds of linguistic issues.

- **Penalty for new words**: LVCSR vocabulary can be modified with new words or terms as they appear. But for that, the entire archive file needs to be processed. Also, as time progresses, a huge number of new words are introduced in the language. It is really a costly affair, both in time and space to continuously update the vocabulary. In case of phonetic search, the dictionary is consulted during the search phase. Addition of new words involves only one extra search. Most of the cases are handled automatically by the spelling-to-sound database, and hence, addition of new words often becomes irrelevant.

- **Phonetic and inexact spelling**: Sometimes, common nouns, e.g., names, can have different spelling. For example, the world's third highest mountain "Kangchenjunga" could be also spelled as "Kangchen junga". LVCSR system requires all the words should be present in the vocabulary, so in this case, the search fails. In phonetic search, exact spelling is not required.

- **User-determined depth of search**: If a particular word is not uttered clearly or there is some background noise, then LVCSR system might not recognize the word correctly. The phonetic search could return multiple results sorted by the individual confidence level. If the search is very important or there is enough time, then one motivated user may search through all possible options until the desired word is not found. Therefore, in phonetic search, user can drill down as deeply as possible.

- **Parallel execution**: The parallel processing paradigm could be utilized in the phonetic searching architecture to increase its performance. For example, the speed of preprocessing could nearly become double if dual processors are used. Bank of computers could be used for parallel processing of the phonetic search tracks to increase the searching speed.

1.6 Summary

We continuously get large quantities of information, both audio and video, in our daily lives in today's information technology age; hence, there is a growing need for efficient ways of searching and retrieving relevant information. Indexing is a process to organize the data according to a specific schema or plan to make the information more presentable and accessible. The main aim of an audio indexing system is to provide the capability of searching and browsing through audio content. Information retrieval methods, e.g., large vocabulary continuous speech recognition [14–17] or phoneme-based speech recognition, are integrated to form an efficient content-based audio indexing system. LVCSR system first generates a textual transcription of the audio content, and thereafter, information retrieval (IR) algorithms are used to index the resulting textual documents [18]. The index then could be used to retrieve relevant portions of the audio documents using standard word query terms. The phoneme-based indexing system works only with sounds. All words are sets of phonemes. In this process, first sounds in a piece of audio content are analyzed to create a phonetic-based index. Thereafter, the user's search term is converted into correct phoneme string by searching through a dictionary of several phonemes. The system then looks for the term in the search query.

In LVCSR system, the word should be present in the vocabulary, so it is dependent on the dictionary. Since audio content is converted into textual representation and textual mapping in done while audio searching, LVCSR is both fast and accurate.

In phoneme-based search, the search key is broken down into individual phonemes, and phonemes are matched in the dictionary. The search term can be effectively predicted in this method. Since this is a probabilistic method for identifying a word so it does not suffer from out-of-vocabulary problem. The phoneme-based search method though can generate false matches.

References

1. Gaikwad, B. M. G. P. *Different indexing techniques*.
2. Foote, J. (1999). An overview of audio information retrieval. *Multimedia Systems, 7*(1), 2–10.
3. Müller, M. (2015). Content-based audio retrieval. In *Fundamentals of music processing* (pp. 355–413). Cham: Springer.
4. Leavitt, N. (2002). Let's hear it for audio mining. *Computer, 35*(10), 23–25.
5. Mand, M. K., & Nagpal, D. (2013). Gunjan, "An Analytical Approach for Mining Audio Signals". *International Journal of Advanced Research in Computer and Communication Engineering, 2*(9).
6. Retrieved September 06, 2018, from http://www.rsystems.com/CommonResource/ KnowledgeRepository/Deciphering-Voice-of-Customer-through-Speech-Analytics.pdf.
7. Logan, B., Goddeau, D., & Van Thong, J. M. (2005, March). Real-world audio indexing systems. In *IEEE International Conference on Acoustics, Speech, and Signal Processing, 2005. Proceedings. (ICASSP 2005)* (Vol. 5, pp. v-1001). IEEE.
8. Cardillo, P. S., Clements, M., & Miller, M. S. (2002). Phonetic searching vs. LVCSR: How to find what you really want in audio archives. *International Journal of Speech Technology, 5*(1), 9–22.
9. Van Thong, J. M., Goddeau, D., Litvinova, A., Logan, B., Moreno, P., & Swain, M. (2000, April). Speechbot: A speech recognition based audio indexing system for the web. In *Content-based multimedia information access-Volume 1* (pp. 106–115). The Centre de Hautes Etudes Internationales d'informatique Documentaire.

10. Van Thong, J. M., Moreno, P. J., Logan, B., Fidler, B., Maffey, K., & Moores, M. (2002). Speechbot: An experimental speech-based search engine for multimedia content on the web. *IEEE Transactions on Multimedia, 4*(1), 88–96.
11. Logan, B., Moreno, P., Thong, J. M. V., & Whittaker, E. (2000). An experimental study of an audio indexing system for the web. In *Sixth International Conference on Spoken Language Processing*.
12. Trieu, H. L., Nguyen, L. M., & Nguyen, P. T. (2016). Dealing with out-of-vocabulary problem in sentence alignment using word similarity. In *Proceedings of the 30th Pacific Asia Conference on Language, Information and Computation: Oral Papers* (pp. 259–266).
13. Yusnita, M. A., Paulraj, M. P., Yaacob, S., Bakar, S. A., Saidatul, A., & Abdullah, A. N. (2011, March). Phoneme-based or isolated-word modeling speech recognition system? An overview. In *2011 IEEE 7th International Colloquium on Signal Processing and its Applications (CSPA)* (pp. 304–309). IEEE.
14. Dey, N., & Ashour, A. S. (2018). Challenges and future perspectives in speech-sources direction of arrival estimation and localization. In *Direction of arrival estimation and localization of multi-speech sources* (pp. 49–52). Cham: Springer.
15. Dey, N., & Ashour, A. S. (2018). *Direction of arrival estimation and localization of multi-speech sources*. Springer International Publishing.
16. Dey, N., & Ashour, A. S. (2018). Applied examples and applications of localization and tracking problem of multiple speech sources. In *Direction of arrival estimation and localization of multi-speech sources* (pp. 35–48). Cham: Springer.
17. Dey, N., & Ashour, A. S. (2018). Microphone array principles. In *Direction of arrival estimation and localization of multi-speech sources* (pp. 5–22). Cham: Springer.
18. Karaa, W. B. A., & Dey, N. (2017). *Mining multimedia documents*. CRC Press.

Chapter 2
Speech Processing and Recognition System

2.1 Introduction

In the initial decade of the twentieth century, scientists in the Bell System realized that the idea of universal services like telephony services is becoming feasible due to large-scale technological revolution [1]. The concept of connecting any telephone users among each other without the need of third party operator assistance became a major challenge and vision for the future of telecommunications. After performing several researches and overcoming the technical challenges, finally by the end of 1915, experimental setup of the first automatic long-distance (transcontinental) telephone call was successfully done. After that, within a few years, the dream of achieving universal telecommunication became reality in the Unites States. Another revolution that has been happened regarding communications was multimedia communication. It was started from the last decade of twentieth century. The vision was to provide continuous seamless, easy, reliable, and high-quality communication between people and machines. There are several ways through which a human can interact with a machine, e.g., text, graphical, touch screen, mouse, speech, etc. It could be argued that for the majority of the population, speech is the most intuitive and natural communication method. Speech is closely related to the language and linguistics falls under social science. Human physiological activity is dependent on speech. It is also related to sound and acoustics. One of the most stirring signals that people work with each and every day is speech.

Speech processing is a special case of digital signal processing (DSP) which is applied to process and analyze speech signals [2–6]. Some of the typical applications of speech processing are as follows: speech recognition, speech coding, speaker authentication, speech enhancement, that is, detection and removal of noise, speech synthesis, that is, text to speech conversion. In order to perform speech recognition, there is a need to understand human speech production system which is described briefly in Sect. 2.2. This section gives a vivid description of human speech production system and corresponding method of establishing an electrical speech production system. This will help in understanding the basics of artificial speech generation system. In Sect. 2.3 a, the basic idea about automated speech recognition (ASR) system is given. The first step of an ASR system is speech to phoneme mapping. Hidden Markov model (HMM) [7] was the most popular model till 1980s. It is described in Sect. 2.3.2. In recent years, there is a revolution in speech recognition with the advent of neural network and deep learning system. This is elaborated in Sect. 2.3.3. After recognition of phonemes, the next step is phoneme to word mapping which is illustrated in Sect. 2.3.4. Section 2.3.5 describes the language model, which deals with arranging the word sequence to form a meaningful sentence. A centralized decoder is responsible to search the correct sentence from the large pool of identifying sentences. This is described in Sect. 2.3.6.

2.2 Human Speech Production System

Human speech system consists of the lungs, trachea (windpipe), larynx, pharyngeal cavity (throat), oral or buccal cavity (mouth), and nasal cavity (nasal tract) [8]. The pharyngeal and the oral cavity are combined into the oral tract. These organs collaborate with each other to generate the speech. In the below sections, speech generation system and corresponding electrical model are described step by step.

2.2.1 Speech Generation

The first stage of the speech production system is message formulation. Before speaking, the speaker constructs the message in his/her by thought process which would be transmitted to the listeners. The formulated message is then converted to a language code. It consists of a set of phonemes which are the smallest units of sounds to form the word. Along with this, duration, loudness, and pitch of the sound are taken care of. After choosing the language code, a series of neuromuscular commands are executed by the speaker to vibrate the vocal cords and to shape the vocal tract in such a way that proper sound sequences are created. The final output of this phase is an acoustic waveform. From Fig. 2.1, it could be observed that muscle force is applied so that air can come out from the lungs through the larynx. After that, a quasi-periodic wave is generated due to the vocal cord's vibration and interruption of the air. The pressure impulse stimulates the air in the oral tract and for the certain sounds also the nasal tract. A sound wave (speech signal) is radiated when the cavities resonate. Vocal and nasal tracts act as resonators having corresponding resonance frequencies.

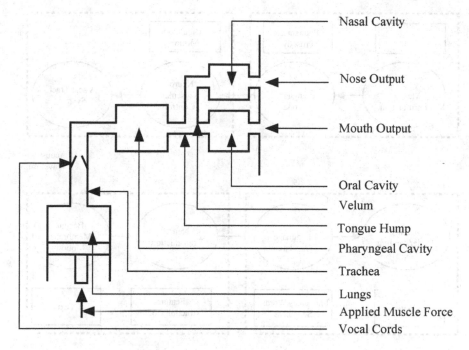

Nasal Cavity

Nose Output

Mouth Output

Oral Cavity
Velum
Tongue Hump
Pharyngeal Cavity
Trachea
Lungs
Applied Muscle Force
Vocal Cords

Fig. 2.1 Block diagram of human speech production system [9]

2.2.2 Speech Perception

The speech perception process begins after the speech is generated and propagated
to the listener. The spectral signal of the acoustic waveform is analyzed at the basilar
membrane which converts the same as activity signals on the auditory nerve. Different
audio features are extracted during this phase. At the higher center of processing of
Human brain, this auditory nerve neural activity is converted into suitable language
code. Finally, after all these stages, the meaning of the message is formed. In Fig. 2.2,
the block diagram of the abovementioned points regarding speech generation and
speech perception process is shown.

2.2.3 Voiced and Unvoiced Speech

There are two types of speech: voiced and unvoiced. The usage of vocal tract and
vocal cords distinguishes between them. In case of voiced sound, vocal cord and vocal
tract are used. The vocal cord vibration produces different voiced sounds and due to
this, the fundamental frequency of the speech could be obtained. While pronouncing,
unvoiced sound vocal tracts are not used; as a result, fundamental frequency could

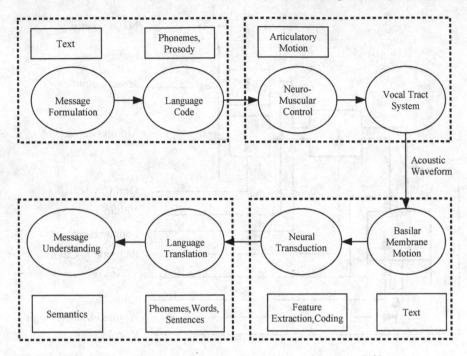

Fig. 2.2 Block diagram of human speech production and perception system [10]

not be obtained. In general, all vowels are of voiced sounds. Sounds like "sh", "p", and "s", etc., are examples of unvoiced sound. Different methods are used to detect whether a sound is voiced or unvoiced. Fundamental frequency could be used as a decision parameter. Another way to detect a voiced or unvoiced sound is by calculating and comparing energy in the signal frame. An unvoiced sound has significantly lesser energy than a voiced sound.

2.2.4 Model of Human Speech

The supralaryngeal vocal tract, which consists of both the nasal and oral cavity, acts as a time-varying low-pass filter (LPF) that suppresses sounds at certain frequencies while allowing its passage at some other frequencies. Transfer function for this LPF could be determined by the whole vocal tract, which is serving as an acoustically resonant system. The model must decide whether the speech is voiced or unvoiced. A pitch detector [11–13] does this. Pitch detector also can extract the fundamental frequency, which helps to control the impulse train. Figure 2.3 shows the model of human speech and the corresponding speech mode of an electrical system.

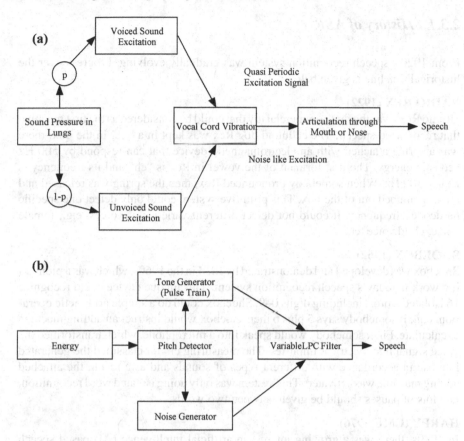

Fig. 2.3 **a** Model of the human speech production system and **b** equivalent speech model for an electrical system

2.3 Automatic Speech Recognition System

The methodology by which the spoken words are converted into textual representation, by the use of computer and software-based techniques to recognize and process human voice, is called as automatic speech recognition (ASR) [14]. The well-known examples of ASR systems are dictation system, voicemail transcription, YouTube close captioning, Google voice search, etc. ASR system is strictly associated with speech translation but not with the speech understanding [15]. If someone instructs a machine by saying, for example, "Urgently Call me a doctor", instead of calling a doctor the response comes as "From now on I will call you a doctor"; it is not the fault of the ASR system, and it is the fault of the speech understanding module. ASR simply concentrates on the conversion of the speech to text.

2.3.1 History of ASR

From 1920s, speech recognition system was gradually evolving. Progress over the historical timeline is given below.

RADIO REX (1922)
"Radio Rex" was the first commercial toy that could be considered as the first machine that recognized speech. The celluloid dog Rex was kept in a box. In the box, there was a spring attached with an electromagnetic device that can respond by 500 Hz acoustic energy. The first formant of the vowel in Rex is "eh" and its frequency is nearly 500 Hz. When somebody pronounced Rex, then the spring was released and the dog jumped out of the box. This primitive system could only detect one specific hardcoded frequency. It could not detect different varieties of voices, e.g., female voice, child voice, etc.

SHOEBOX (1962)
Shoebox was developed and demonstrated by IBM in the 1960s, which was a pioneering work in today's speech recognition system. This unique device could recognize 16 isolated words, including digits 0–9. Shoebox could do a simple arithmetic operation, e.g., if somebody says 5 plus 6 then shoebox would instruct an adding machine to calculate. First, somebody would speak into a microphone, which transformed the voice signal into electrical impulses. The measuring circuit classified the generated impulse in accordance with different types of sounds and as a result, the attached adding machine was activated. This system was only doing isolated word recognition, i.e., lots of pauses should be given between two words.

HARPY (CMU 1976)
In 1970s, there was a growing interest in artificial intelligence (AI) based speech recognition system. This was a three million dollar project taken by ARPHA. The aim was to develop an advanced speech recognition system that could identify continuous vocabulary. Finally, HARPY was invented from CMU in 1976. It could recognize a vocabulary containing 1000 words. It is also having a limited vocabulary and it did not use any statistical method. It was purely rule-based.

Statistical Model-Based System (1980)
The statistical models were becoming popular from the period of 1980s, and the most frequently used model regarding this was hidden Markov model. The previous models were all rule-based. The statistical models generalized the speech recognition system. The vocabulary size was getting larger and near about 10 k vocabulary was used. This was also called large vocabulary continuous speech recognition (LVCSR).

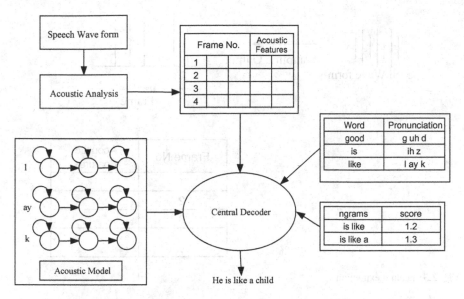

Fig. 2.4   Structure of an ASR system [15]

Deep Neural Network-Based Speech Recognition System (From 2006 Onwards)
In 2006, deep neural network was introduced. Nowadays, each and every speech recognition system uses deep neural network, e.g., SIRI, Cortana, Google Voice Search, etc.

2.3.2   Structure of an ASR System

The design and development of a speech recognition system are dependent on various components [15] like representation and preprocessing, speech classes, different types of feature extraction techniques, classifiers used, database, and performance of the system. Figure 2.4 represents the typical structure of an ASR system.

(i)   **Acoustic Analysis**

In this stage, the input sound waveform is converted into some discrete form or feature vectors. Normally, a speech signal is not stationary but when a short time span near about 20 or 25 ms is considered, then it acts as a stationary signal. This short discrete signal snap is called a frame. The acoustic model extracts features [16, 17] from frames. Feature extraction deals with obtaining the useful information of a signal by removing the irrelevant information. The main aim of the feature extraction in ASR system is to find some set of feature vectors or properties that can give a compact representation of the input audio signal. Acoustic model is the next stage of the ASR, where a statistical model is found with the help of the extracted feature

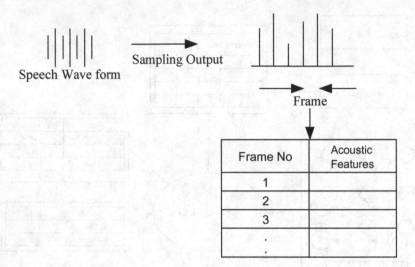

Fig. 2.5 Feature extraction

vectors. These extracted features are input to the next stage of the ASR that is the Acoustic model, where a statistical model is found with the help of feature vectors. Figure 2.5 shows the feature extraction process from an audio frame.

(a) **Phoneme as basic units of Acoustic information**

The smallest distinct or distinguishing part of a language is called a phoneme, e.g., five can be broken down into phonemes like "f", "ay", "v". Transforming a spoken word into individual phonemes is done by the language experts. Each and every language has 50–60 phonemes on average. Word could not be used as the basic unit of speech recognition because it is not possible to construct a vocabulary having all combinations of words. If a word which is not in the training vocabulary set appears, then the speech recognition system fails to identify the word. For example, if our vocabulary contains following three words: "five", "four", and "one", respectively, if a new word "nine" comes, then it could not be identified. Now in phonetic, representation "five", "four", and "one" can be broken down into following individual phonemes:

Five → "f", "ay", "v".
Four → "f", "ow", "r".
One → "f", "ah", "n".

If these phonemes are already in the vocabulary, then if a new word "nine" comes, it could be identified by the existing phonemes as follows:

Nine → "n", "ay", "n"

(ii) **Acoustic model: Acoustic features to phoneme transformation**

The relationship between an audio signal and phonemes is represented by an acoustic model. A word is made up with some distinct sounds which are called phonemes.

Fig. 2.6 Markov chain
showing the probability of a
sequence of weather events

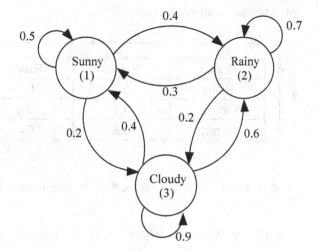

Phonemes act as a basic unit of speech. The statistical representation of each of these distinct sounds is kept in a file. A speech corpus or speech database is created and some training algorithms are applied to it to create a statistical representation for each phoneme. These statistical representations of each phoneme are called hidden Markov model (HMM). A deep insight of HMM is given in the subsequent section.

(a) Hidden Markov Model (HMM)

The Hidden Markov Model is one of the most important machine learning models in speech and language processing. It is the extension of the Markov model. Markov model or Markov chain is based on weighted finite state automata. Here, each arc in the state transition is associated with a weight which is the transition probability from one state to another state. Markov model is a fully observable model [18]. The state transitions have been observed for some time and state transition probabilities are calculated based on these observations. The probability of all the arcs leaving from a node (individual states are considered as nodes) should be 1. The state transition is determined by the input sequence. Figure 2.6 represents a Markov model for assigning probabilities to the weather state transitions. A Markov chain consists of the following:

$$Q = q_1 q_2 \ldots q_n - \text{a set of N states}$$
$$P = p_{01} p_{02} \cdots p_{n1} \cdots p_{nn}$$

where p_{ij} is the probability of moving from state i to state j, and $\sum_{j=1}^{j=n} p_{ij} = 1$.
   A first-order discrete-time Markov chain only depends on the previous state.
   If at time t the current state is $q_t = j$, the probability of state q_t is

$$P\big[q_t = j | q_{t-1} = i, q_{t-2} = k, \ldots\big] = P\big[q_t = j | q_{t-1} = i\big]$$

Table 2.1 State characterization table

j

States	Sunny	Rainy	Cloudy
Rainy	0.3	0.7	0.2
Cloudy	0.4	0.6	0.9
Sunny	0.5	0.4	0.2

i

If there are N numbers of states, then the state transition probability is given by

$$p_{ij} = P[q_t = j | q_{t-1} = i] \; where \; 1 < i, j < N \; and \; p_{ij} \geq 0 \left\{ \forall i, j \sum_{j=1}^{j=N} p_{ij} = 1 \right\}$$

(2.1)

The weight on the edges represents the probability of the state in the next day given a particular state in the previous day, e.g., if today is a sunny day the probability of the next day to be rainy day is 0.4. The weather on a particular day t is dependent on the state characterization Table 2.1.

The probability of the weather for eight consecutive days to be "cloudy-cloudy-cloudy-rainy-sunny-sunny-cloudy-sunny" could be found in the following way and represented as observation sequence:

$$O_{seq} = (cloudy, \; cloudy, \; cloudy, \; rainy, \; sunny, \; sunny, \; cloudy, \; sunny\}$$

This could be represented by individual states as $O_{seq} = (3,3,3,2,1,1,3,1\}$.
Now, from Eq. 2.1, the following could be derived:

$$P[O_{seq} | Model] = \prod_{t=1}^{n} p[q_t | q_{t-1}] = p[3]p[3|3]p[3|3]p[2|3]p[1|2]p[1|1]p[3|1]p[1|3]$$

$$= 1 * 0.9 * 0.9 * 0.6 * 0.3 * 0.5 * 0.2 * 0.4 = 0.0058$$

Markov model can only determine fully observable events [18], i.e., it can work only with unambiguous sequences but fails to represent inherent ambiguous problems. It is a process where each state is mapped to a deterministic observable event. Hence, Markov model could not be applied in many practical problems, including automatic speech recognition. The concept of Markov model is extended into hidden Markov model (HMM) [19] in which the modeled system is considered as a Markov process which passes through unobserved (i.e., hidden) states. The resulting HMM is a conglomeration of a stochastic process with another stochastic process which is not observable. This can only be observed through another set of stochastic process

Fig. 2.7 One-coin model

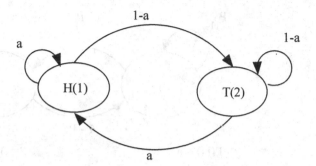

Observation Sequence:H T H H T H H T
Corresponding State: 1 2 1 1 2 1 1 2

which produces the sequence of observations. The abovementioned Markov model and HMM could be explained by **coin-toss** [20] models as below.

Coin-toss model: In this model, it is assumed that two persons are in two rooms separated by a barrier. In one room one person is performing coin-tossing experiments and stating loudly the sequence of head and tails. In the other room, the person can listen to whatever the other person is telling, but cannot observe how many coins have been tossed due to the barrier. So the number of coins tossed is hidden to him. So depending on the number of coins tossed, there are different coin-toss models as mentioned below.

One-coin scenario: If the person tossing one single coin, then the following state diagram (Fig. 2.7) could be formed:

It is assumed that state 1 is when the person gets a head and state 2 is when he gets a tail.

Here, a = probability of getting a head, i.e., a = P(H).

Here, a biased coin is considered, so the probability of getting a tail P(T) = 1 − a.

There is only one unknown, i.e., the probability of getting a head.

Since only one coin is tossed so the corresponding state could be identified from the observed sequence of head and tail, one-coin-toss model is fully observable and falls under a simple Markov model.

Two-coin scenario: Here, the person performing coin-tossing experiment is tossing two coins. Below is the state diagram (Fig. 2.8) for that.

Here, tossing the coin 1 is state-1 and tossing the coin 2 is state-2 and a_1, a_2 are the probabilities of getting a head in state-1 and state-2. There are four probable events in this scenario: coin 1 is tossed and head is obtained, the coin 1 is tossed and tail is obtained, the coin 2 is tossed and head is obtained, and coin 2 is tossed tail is obtained. In this particular case, it is not possible to identify which coin is tossed by simply observing the head and tail sequence. The state is hidden here and it falls under the hidden Markov model. The states could be determined if a third

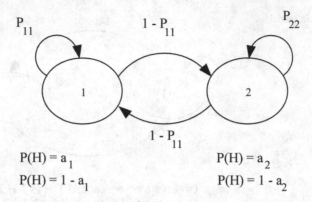

$P(H) = a_1$

$P(H) = 1 - a_1$

$P(H) = a_2$

$P(H) = 1 - a_2$

Fig. 2.8 Two-coin model

Table 2.2 State characterization table showing the probability of carrying an umbrella

	Probability of carrying an umbrella
Cloudy	0.4
Rainy	0.7
Sunny	0.2

coin is used here. After tossing this third coin, if a head is obtained, the person doing the coin-tossing experiment might choose coin 1 or could choose coin 2. Here, the observing person can identify the states based on another stochastic process, i.e., the tossing of the third coin.

(b) **Application of Bayes' Theorem on the HMM problem**

Now, if the previously mentioned weather example is slightly modified as below:

A person is locked inside a room for several days and he is asked about the weather outside. There is a caretaker who comes in the room to provide his meal. The only way to make an assumption of the outside weather is to observe that the person carries an umbrella or not.

The probability of the caretaker carrying an umbrella is given in Table 2.2.

Markov model is fully observable whereas in the abovementioned problem, the actual outside weather is hidden from the person who is inside the room. The assumption of the outside weather should be done with a related event, i.e., whether the caretaker carries an umbrella or not. The next section describes how the probability of the outside weather could be calculated by using Bayes' theorem which describes the probability of an event, based on prior knowledge of conditions that might be related to the event [20].

The Bayes' theorem could be expressed mathematically as

$$P(X|Y) = \frac{P(Y|X)P(X)}{P(Y)} \tag{2.2}$$

Here, X and Y are two events and $P(Y) \neq 0$.

$P(X|Y)$ is the conditional probability: The probability of event X given that the event Y has already happened.

$P(Y|X)$ is also another conditional probability: The probability of event Y given that the event X has already happened.

$P(X)$ and $P(Y)$ are the probability of occurrences of X and Y independently.

From Eq. 2.2, the following could be obtained

$$P[q_1, q_2, \ldots, q_n | o_1, o_2, \ldots, o_n] = \frac{P[o_1, o_2, \ldots, o_n | q_1, q_2, \ldots, q_n] P[q_1, q_2, \ldots, q_n]}{P[o_1, o_2, \ldots, o_n]}.$$

where $o_1, o_2, \ldots, o_n$ are the respective observations and states are $q_1, q_2, \ldots, q_n$.

Here, o_j is true if on a particular day i the caretaker brought the umbrella and false if the caretaker did not bring the umbrella. The probability $P[o_1, o_2, \ldots, o_n]$ is the prior probability to observe a particular umbrella event. $P[o_1, o_2, \ldots, o_n | q_1, q_2, \ldots, q_n]$ could be estimated by using the Bayes' theorem in the following way:

$$P\left[o_1, o_2, \ldots, o_n | q_1, q_2, \ldots, q_n\right] = \prod_{t=1} p[o_t | q_t]$$

$$\left[\forall i \; q_i, \; o_i \text{ is independent of all } o_j, \; q_j \text{ provided } j \neq i\right]$$

Let the person was locked in the room on a sunny day. Now, the problem is "Estimate the weather of the next day is cloudy provided that the caretaker brought the umbrella in the next day."

$$Here \; q_1 = sunny, \; q_2 = cloudy \; o_2 = true \; and \; P(o_2) = 0.5$$

$$Now \; P[q_2 | q_1, o_2] = \frac{P[q_2, q_1 | o_2]}{P[q_1 | o_2]} \left[since \; p(A|B) = \frac{p(A, B)}{P(B)}\right] \quad (2.3)$$

Since q_1 and o_2 are the independent events, o_2 could be ignored in Eq. 2.3, and the same could be written as follows:

$$P[q_2 | q_1, o_2] = \frac{P[q_2, q_1 | o_2]}{P[q_1 | o_2]} = \frac{P[q_2, q_1 | o_2]}{P[q_1]} \quad (2.4)$$

Using Bayes' theorem, Eq. 2.4 could be expanded as

$$\frac{P[q_2, q_1 | o_2]}{P[q_1]} = \frac{1}{p[q_1]} \frac{P[q_2, q_1] P[o_2 | q_2, q_1]}{P[o_2]}$$

Now, the probability of today's umbrella event does not depend on yesterday's weather state, i.e., q_1 here. This is the first-order Markov assumption.

The above expression can be written as

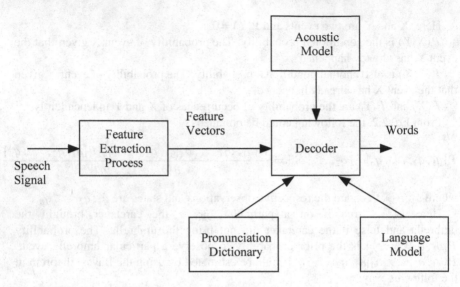

Fig. 2.9 HMM-based LSVCR speech recognizer [21]

$$\frac{1}{p[q_1]}\frac{P[q_2,q_1]P[o_2|q_2,q_1]}{P[o_2]} = \frac{1}{P[q_1]}\frac{P[q_2,q_1]P[o_2|q_2]}{P[o_2]} = P[q_2,q_1]\frac{P[o_2|q_2]}{P[q_1]P[o_2]}$$

$$= P[q_2|q_1]P[q_1]\frac{P[o_2|q_2]}{P[q_1]P[o_2]} \; (since \; P[A,B] = P[A|B]P[B])$$

$$= P[q_2|q_1]\frac{P[o_2|q_2]}{P[o_2]}$$

Now, the probability that the next day is cloudy given that the previous day is sunny and the caretaker has brought the umbrella could be calculated by the above formula:

$$P[q_2 = cloudy \,|q_1 = sunny]\frac{P[o_2 = caretaker \; brought \; the \; umbrella \,|\, q_2 = cloudy]}{p[o_2 = caretaker \; brought \; the \; umbrella]}$$

$$= \frac{0.2*0.4}{0.5} = 0.16$$

(c) Application of Hidden Markov Model in speech recognition

In Fig. 2.9, the principle components of a large-scale vocabulary speech recognizer are shown. Initially, a sequence of fixed-sized acoustic vectors, $V_{1:T} = V_1, V_2 \ldots, V_T$, are generated from the input waveform which is called feature extraction. After that, the decoder tries to find the probable word sequence $W_{1:L} = W_1, W_2, W_3, \ldots, W_L$ which is most likely responsible to generate V.

Here, the observations are the speech vectors and the hidden states are word sequences.

The same could be mathematically written as

$$W = arg\ max\{P(W|V)\}\ where\ W \subseteq L\ and\ L\ is\ the\ language\ in\ concern$$

The decoder tries to find the word sequence which has the maximum probability to generate the acoustic vector. Now, it is difficult to model $(W|V)$. If Bayes' rule is applied to the above equation, then

$$W = arg\ max\left\{\frac{P(V|W)P(W)}{P(V)}\right\}$$

For a single speech input, P(V) is constant. So ultimately, the above problem concentrates on finding

$$arg\ max\{P(V|W)P(W)\}$$

$P(V|W)$ is determined by an acoustic model and prior probability P(W) is determined by a language model. For any given W the associated acoustic model is generated by concatenating phones to construct words as defined by a pronunciation dictionary. The parameter for this model could be estimated from the training data consisting of speech waveforms and their corresponding orthographic transcriptions. The language model is an N-gram model where the probability of a word depends on the previous n words. The decoder in this way searches for all possible word sequence and prunes the irrelevant hypotheses and as a result, the most likely word sequence is obtained.

(d) **Limitation of HMM in Speech Recognition**

Hidden Markov Model in Speech Recognition has its inherent limitations [22]. It is assumed that the successive observations are mutually independent but in reality, those are mostly dependent on each other. HMM is based on the Markov property. The basic assumption of Markov model is that the probability of a given state at a particular time t is only dependent on the state at time t−1. But in case of speech sounds it is observed, that dependencies frequently extend through more than one state. Moreover, the frame size is constant here. There is no well-defined formal method for deciding the overall architecture to solve a problem. Also, there is no specific method for deciding the number of states and transitions needed for a model. In HMM, a huge number of decision parameters are needed and the amount of data required to train an HMM is very large.

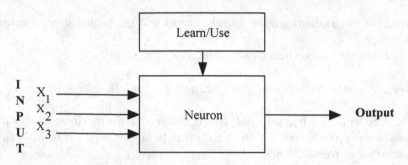

Fig. 2.10 A simple neuron

Table 2.3 Truth table before applying the firing rule

P	0	0	0	0	1	1	1	1
Q	0	0	1	1	0	0	1	1
R	0	1	0	1	0	1	0	1
OUT	0	0/1	0	0/1	0/1	0/1	1	1

2.3.3 Neural Network and Speech Recognition System

An artificial neural network (ANN) is an information processing system that closely follows the way biological nervous system functions (e.g., working mechanism of brain). A huge number of interconnected processing nodes or systems (neurons) are working together to solve specific problems. Such as human beings learn from the past, ANN also learns in the same way. Figure 2.10 shows a simple neuron. Implementation of ANN [23–26] deals with a huge number of inputs and one output. Here, the neurons undergo two modes of operations. First, the neurons are subjected to training mode. In this mode, the neurons are trained to detect a particular input pattern. After that, the neurons are put into the execution mode. When a previously taught input pattern appears into the input, the associated output becomes the current output. If the input pattern does not belong to the previously taught list, the firing rule is used to know the output.

A firing rule is developed to determine whether a neuron should fire or not for a particular input pattern. It is not specific to the taught input pattern, but applies to all possible input patterns.

An example related to a three input neuron system is given below. When the input (P, Q, and R) is 111 or 110, then the output is 1 and when the input is 000 or 010, then the output is zero. Before applying the firing rule, the truth table is given in Table 2.3.

Hamming distance is considered to construct the firing rule. Considering the pattern 001, the distance between 000 and 100 is of 1 bit, 010 and 100 is of two bits, 110 and 100 is of 1 bit, and 111 and 100 is of two bits. So 010 have an equal chance of being 1 and 0. The specific output could not be decided in this case. If the pattern

Table 2.4 Truth table after applying the firing rule

P	0	0	0	0	1	1	1	1
Q	0	0	1	1	0	0	1	1
R	0	1	0	1	0	1	0	1
OUT	0	0	0	1	0/1	1	1	1

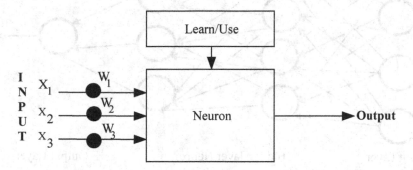

Fig. 2.11 A MCP neuron

001 is compared, it is observed that it is nearest to 000. So the output is zero here. In this way, the firing rule could be constructed and the corresponding truth table is shown in Table 2.4.

(i) McCulloch and Pitts Model (MCP)

The abovementioned neural network is a very basic model and does not perform anything better than a conventional computing system. In Fig. 2.11, a more advanced version of the previous basic neural model is shown, which is known as "McCulloch and Pitts model (MCP)". Here, the inputs are weighted. The final decision-making is dependent on the particular weighted input. After adding the weighted inputs the resulting weighted sum is compared with a predefined threshold value and if it is greater than that then the neuron fires otherwise not.

So here the neuron fires if and only if:

$$X_1 W_1 + X_2 W_2 + X_3 W_3 + \ldots X_n W_n > T$$

where T is the threshold value; $X_1, X_2, \ldots, X_n$ are the input vectors; and $W_1, W_2, \ldots, W_n$ are the corresponding weights.

This neuron is a very flexible and powerful one because of the introduction of threshold value and input weights. It is adaptable to the changing environment by changing the input weights and the threshold.

(ii) Types of Neural Networks:

There are two types of neural networks: feedforward network and feedback networks or recurrent neural network (RNN).

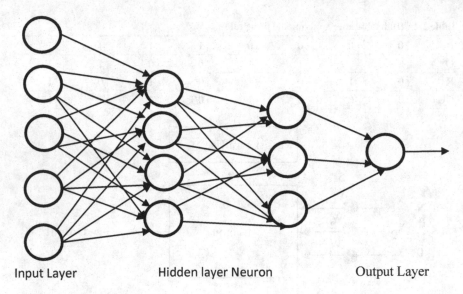

Input Layer   Hidden layer Neuron   Output Layer

Fig. 2.12 A feedforward network

(a) **Feed Forward Networks**: In Fig. 2.12 a, simple feedforward network is shown. The signal is allowed to travel in one way only. In this network, the leftmost layer is the input layer containing the input neurons, and the rightmost layer contains the output neurons. One or more hidden layers are there in the middle of the input and output neurons. The functionality of the hidden layers is to transform the inputs for the use of the output layer. For example, a computer system is assigned to detect the presence of a bus. The bus detector tool could be thought of as a combination of several unit layers like a layer for detecting the wheels, a layer for calculating the size, and a layer for detecting the shape. These are the hidden layers and they are not part of a raw bus image. These layers only help to identify the different parts of a bus and collectively these will be used to recognize the whole bus.

If x, y, and h represent the input vector, hidden layer activation, and the resulting output vectors, respectively, then the above could be represented as follows:

$$h = f(x) \ where \ f \ is \ a \ function \ that \ maps \ x \ to \ h$$

$$y = g(h) = g(f(x)) \ where \ g \ is \ a function \ that \ maps \ h \ to \ y$$

(b) **Feedback Neural Network**: In a feedback neural network, signals are allowed to travel in both directions, thus introducing loop in the resulting network. Feedback networks are very dynamic and powerful. Their status changes continuously until equilibrium is reached. A new equilibrium is found when the input changes. Feedback architecture is also called recurrent neural network (RNN). This kind of network is most suitable for time series prediction. After transferring the input data to a network

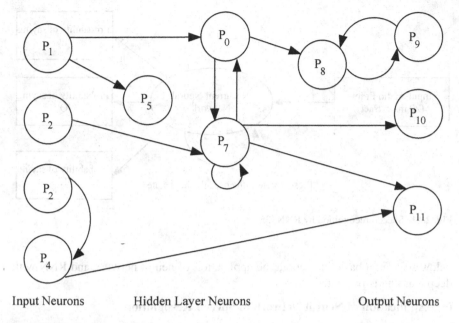

Input Neurons   Hidden Layer Neurons   Output Neurons

Fig. 2.13 Feedback network

element (hidden layer), it is memorized and could be fed into the subsequent inputs. Hence, past data could be utilized to predict the current and as well as the future states. In Fig. 2.13, a simple feedback neural network is shown.

(iii) **Deep Learning and Speech Recognition**

Deep neural networks [27–31] differ from the single-hidden layer neural networks by the number of intermediate hidden node layers. A neural network having three or more layers including the input and output layers qualify for the deep learning network. In deep neural network, each layer of the node trains from the previous layer output. Thus, traversing through the subsequent layers of a deep neural network means recognizing the more complex features as it aggregates the features from the previous layers. In the previous section, a basic idea of the RNN is given. Recurrent network takes the current input and also previous perception of the input. Feedback loops in multiple hidden layers can increase the prediction accuracy to a great extent. Geoffrey Hinton is a pioneer in the field of artificial neural networks and copublished his research work on the backpropagation algorithm [32] for training of multilayer perception networks. Since these inner layers are doing some inherent processing which is not visible to the outside world he may have started the introduction of the phrasing "deep" to describe the development of large artificial neural networks. In another research paper titled "Acoustic Modeling using Deep Belief Networks" [33], he showed that better phone recognition could be achieved by replacing HMM with the deep neural network. Furthermore, in a recent research work [34], Geoffrey analyzed the effect of deep recurrent neural network in speech recognition. In the

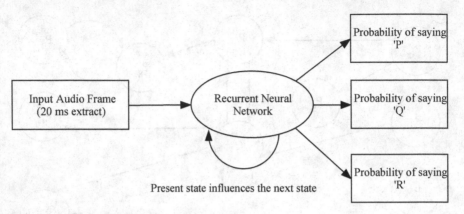

Fig. 2.14 Speech recognition by RNN [35]

below section, a basic idea about the application of neural network and RNN with deep learning is presented.

(a) **Application of Neural Network in Speech Recognition**

Recurring neural network concept could be effectively applied for automatic speech recognition [35]. First, the input audio speech is sampled. This sampled data could be fed directly into the neural network but then it would be very difficult to recognize the speech patterns. In order to make this process easier a certain kind of preprocessing is done on the sampled data. Initially, the sampled data is grouped into 20-millisecond long chunks. In this short time span also, some low, mid, and high pitched sounds are intermingled with each other. Fourier transform is applied on this complex sound part and it is broken into the individual simple component parts, i.e., the entire 20-ms frame is broken into low-pitched part, next low-pitched part, etc. This is done by applying Fourier transformation. After that, the energy contained in each of the frequency bands is added, and thus, a fingerprint for the audio snippet is created. The same process is repeated on each and every 20-ms audio chunks and a spectrogram could be formed out of this. A spectrogram is very useful because one can clearly observe the pitch patterns of the audio data. Compared to raw audio data a neural network can produce pattern more easily in this type of representation. So this audio data representation is the input to the neural network. 20-ms audio chunks are fed into the neural network. It will try to find out the letter corresponding to the spoken sound in each audio slice.

Recurrent neural network has a memory and the future predictions are influenced by it (Fig. 2.14). In speech recognition, the prediction of the next letter depends on the previously predicted letter, e.g., if "hel" is already spoken, then there is a high chance that the next utterance would be more likely "lo" so that the resulting word is "hello". The existence of previous memory in RNN helps the prediction by the neural network, more accurate. In this way, the entire audio is fed into the RNN and each audio chunk is mapped to the most likely spoken word.

(b) **Anomaly resolution by RNN in ASR**

More than one prediction for the same spoken word could be possible and RNN should filter out the correct one for them. If the spoken word is "Hello", then there might be more than one prediction for that, e.g., "HHHHEE LL LLLOOO" or "HH-HHUU LL LLLOOO" or at worst case "AAAUU LL LLLOOO".

To remove this anomaly the following steps are done:

First, the repeated characters are removed. So above three predictions are transformed into:

"HE L LO", "HU L LO", and "AU L LO".

Then, the blank spaces are removed and the following words are found:

"HELLO", "HULLO", and "AULLO".

Now, this pronunciation-based prediction is combined with likelihood scores based on a large database of written text. Based on the likelihood score the most realistic prediction is accepted by the prediction system. Here, "Hello" will have the greatest likelihood score and hence it should be accepted.

2.3.4 *Pronunciation Model*

The output of the statistical model or hidden layers of deep neural network is a set of phonemes associated with their probability measures. The mapping of these phonemes to the specific words is done by pronunciation model. A simple dictionary of a large number of pronunciations is maintained. It basically consists of words and corresponding phoneme sequence mapping.

$$e.g.\ Can \rightarrow Ka + en$$

$$Five \rightarrow f + ay + v$$

$$Four \rightarrow f + ow + r$$

Pronunciation model is the only model in the ASR system which does not undergo learning phase.

This model is built by linguistic experts. Now, this model has its own restrictions. The actual dictionary pronunciation of the word "probability" is "praa b ax b l iy". In reality, "probability" could be pronounced in the following ways: "p r aa b iy", "p r aa l iy", "p r aa b uw", "p aa b uh b l iy", etc. Also sometimes an extra character might be inserted while pronouncing. As an example, "sense" could be pronounced as "s eh n t s". There are lots of different kinds of pronunciations associated with a particular word. One of the efficient ways to resolve this conflict is using speech articulatory knowledge. Our vocal tract consists of different articulators which move in some specific pattern or fashion to produce different speech sounds. When the speaker is about to speak, the central nervous system sends a gesture which acts

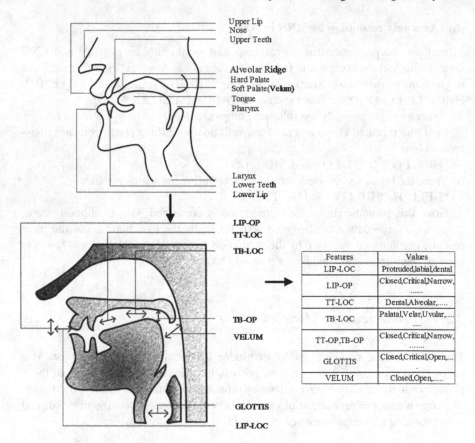

Fig. 2.15 Articulatory variables [37]

as an instruction to the vocal tract to generate the required degree of movement or constriction at a specific location associated with a particular set of articulators. The main articulators are shown in Fig. 2.1a. The automatic tracking of the movement of the articulator's results in accurate identification of the spoken word. Browmen and Goldstein proposed the idea of articulatory phonology [36]. The vocal tract space is discretized and the articulators are treated as individual variables (Fig. 2.15). These variables can take any number of values from 1 to n. The different combinations of these values of the variables produce the specific representations of the spoken words. In this way, sequence of phones could be represented as a stream of articulatory features. This could be utilized to build a pronunciation model [37] for ASR.

2.3.5 Language Model

Pronunciation model helps to organize the phonemes to form a meaningful word. Language model, on the other hand, helps to organize words to form a meaningful sentence. It is basically a statistical model that indicates how likely a sequence of words can appear together. It is useful in many applications especially in natural language processing applications because it is a means to estimate the relative probability of the occurrences of different phrases. The language model is applied in speech recognition, handwriting recognition, parsing machine translation, spelling correction, and many more.

For example, the following word sequences:

The horse "ran" occurs more frequently than "The horse" "can" or "The horse" "Pan". Therefore, the latter two options "can" and "pan" are rejected.

The language model is also responsible to disambiguate between similar acoustics, e.g., if somebody utters "he is famous", it could also map to "he is fame us". Language model searches in the large English vocabulary and finds the relative probability of "he is" with "famous" and "he is" with "fame us". Obviously, the later will be rejected.

Various toolkits have been developed for language modeling. One of the most used toolkits is SRILM toolkit [38]. If someone needs to work with large vocabulary, then there is another toolkit called KenLMtoolkit [39], which is very fast. If someone's work is related to finite state machine, then OpenGram NGram Library developed by Google is a good option. One of the leading language models is the N-Gram model which is discussed in the next section.

(i) N-Gram Model

N-Gram is used to predict the nth word from the previous $(n - 1)$ words, e.g., if the problem is to predict 5th word from the previous 4 words then it is called a pentagram model. If the problem is to find the 4th word from the previous three words, it is called the 4-Gram. Like this one could define bigram, trigram, and so on. For example, "He is going to _____." Here, the fifth word is to be found. It is an example of pentagram model.

(a) Computing probability of a sentence in N-Gram model

If somebody says "I will play football", then there is a possibility of translating it to "Eye will play football" which is not meaningful. Using N-Gram model correct sequence of words could be found by using the following chain rule of probability.

If there are n factors or states $S_1, S_2, S_3 \ldots S_n$, then

$$P(S_1S_2S_3 \ldots S_n) = P(S_1)P(S_2|S_1)P(S_3|S_2S_1)P(S_4|S_3S_2S_1) \ldots P(S_n|S_{n-1} \ldots S_4S_3S_2S_1)$$

For a sequence of words $W_1W_2W_3 \ldots W_n$, the above equation could be written as

$$P(W_1W_2W_3 \ldots W_n) = P(W_1)P(W_2|W_1)P(W_3|W_2W_1)P(W_4|W_3W_2W_1)$$
$$P(W_n|W_{n-1} \ldots W_4W_3W_2)$$

For the sentence "I will play football", the above could be applied as follows:

P (I will play football) = P(I)P(will| I)P(play | I will)P(football | I will play).

In the above example, unnecessary backtracking is required. To eliminate unnecessary going back to several states, Markov assumption could be used, where the main motto is "today is enough for predicting tomorrow … no need to look back days before yesterday."

According to this, the model could be converted into a bigram model.

P (I will play football) = P(I)P(will | I)P(play | will)P(football | play).

The above conditional probabilities could be calculated by analyzing a large vocabulary corpus or training data.

The N-Gram model depends on the training corpus. It might happen that some N-Gram remains unseen. Even after analyzing large corpus data, there might be many unseen bigram/trigrams. For any unseen data, the probability is considered to be as zero. This is also called unsmoothed N-Gram. Here, the probability of observed N-Gram is maximized by assuming the probability of the unseen N-Gram as zero. This zero bigram probability results into errors in speech recognition and results in poor speech recognition as it disallows the bigram regardless of how informative the acoustic signal is. This leads to inaccurate speech recognition. Some smoothing techniques (e.g., interpolation technique, adjustment of probability estimates) have been proposed by Chen and Goodman [40]. These are briefly described below.

(ii) **Smoothing techniques for minimizing unseen N-Gram problem**

Smoothing refers to the adjusting the probability estimate to increase the accuracy level of the speech recognition [40, 41]. One simple technique is to increase the calculated probability count by one. As a result, each bigram occurs minimum once. However, this scheme suffers from the disadvantage of assigning the same probability to the less important or less frequent bigram data, e.g., it might assign the same probability to "I am" and "I is" which is incorrect.

In order to resolve this, an interpolation technique was applied considering the unigram model. Here, likelihood of each word in a bigram is considered. In Bigrams more frequent words are assigned higher probability value.

2.3.6 Central Decoder

In the above sections, all the individual components of an ASR system is discussed. Now, the output of all the models, e.g., words with their likelihood probabilities, etc., is fed to the decoder. The decoder then identifies the most accurate transcription of the spoken sentence. From the decoder perspective, it is mainly a search problem. Considering one such system which only has to recognize one or nine, each has corresponding phonemes and each phoneme have corresponding HMMs. Combining all these components, a search graph (Fig. 2.16) is formed.

From start, one can move to one or nine. It could be observed that for these two words quite a large search graph is obtained. One system typically has thousands of

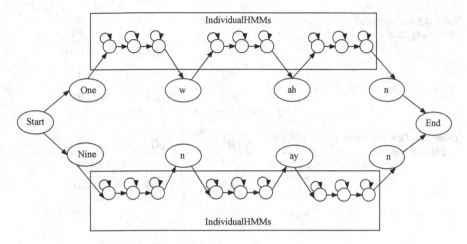

Fig. 2.16 Search graph for one and nine

words and each having a corresponding search graph. The decoder has to search from this huge search space and filter out the most accurate translation. If the vocabulary size is V and there is a sequence of W words, then there are V^W alternative paths in the search graph [42]. There is an exponential growth of different search paths with the increasing number of words in the sentence. One of the pioneering algorithms for this huge search space optimization is the Viterbi algorithm which is discussed in the next section.

(i) **Viterbi algorithm and HMM**

The Viterbi algorithm was proposed by Andrew Viterbi in 1967 as a means of decoding convolutional codes or error codes over noisy channel [43]. It is a dynamic programming algorithm for obtaining the most probable sequences of hidden states (known as Viterbi path) given a sequence of observed events which is especially useful for an HMM problem [44]. This algorithm computes a metric or discrepancy for each and every possible path in the HMM state transition diagram. Among numerous paths ending at a particular state, the winning path is selected based on the metric and retained. The other paths are rejected. This drastically reduces the exponential search space and transfers it to a linear one. Viterbi algorithm is described below in brief.

Viterbi Method

Algorithm:

 (I) Initialization: The starting state is labeled as Q_0.
 (II) Computation: Let j be the number of steps ranging from 0, 1, 2 . . . n. At any previous step j, the following steps are done:

   (a) All possible survivor paths are identified.

Table 2.5 Transitional
probability matrix

	U_1	U_2	U_3
U_1	0.2	0.5	0.6
U_2	0.7	0.3	0.3
U_3	0.4	0.5	0.4

Table 2.6 Observation/output
probability matrix

	R	G	B
U_1	0.4	0.6	0.3
U_2	0.2	0.5	0.6
U_3	0.7	0.2	0.4

    (b)   Survivor path and its metric (here the transition probability) for each state is calculated.

(III) At label j + 1 metric for each path entering each individual state is computed.

(IV) Identify and retain the path having the winning metric, i.e., highest transition probability.

(V) Final Step: Continue step II until the algorithm complete this forward search throughout all the states in the state transaction diagram.

Example:

Suppose there are three urns, U_1, U_2 *and* U_3 containing red (R), green (G), and blue (B) balls.

The probability of moving from one urn to another is given below in the transition probability matrix (Table 2.5). Here, the observation probability is the probability of drawing a particular ball, i.e., red, green or blue from a given urn. The same is given in the observation or output probability matrix (Table 2.6).

The given observation sequence is RRBGGBR. The problem is to find the corresponding hidden state transition path. The state transition diagram (Fig. 2.17) for Tables 2.5 and 2.6 is given in the following.

By Markov chain rule, the state transition probability and the probability of getting red, blue, or green ball can be combined simply multiplying the corresponding probability values. After doing this, the following state transition diagram (Fig. 2.18) is obtained.

For the first symbol in the observation sequence, i.e., "R" the Viterbi algorithm is applied as follows.

First, the state diagram is obtained from the state transition table (Fig. 2.18). It is a tree-like structure assuming the starting state as Q_0. The first symbol in the observation sequence could be considered as a null string or ϵ. From the starting state one could go to state and for that, corresponding observation is assumed as null string or ϵ. Assuming the probability of going to state U_1, U_2 *and* U_3 as 0.6,

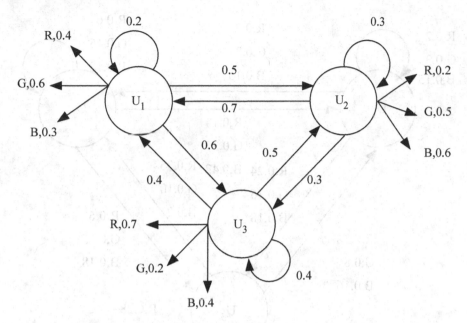

Fig. 2.17 State transition diagram

0.4, and 0.3 from Q_0 for the observation "$\mathfrak{C}$", the following state transition sequence (Fig. 2.19) could be obtained.

From the Markov chain rule, probabilities of the subsequent ages could be multiplied to get the final probability at the leaf node. Among all the paths, the winning paths (shown as sequence of bold edges in Fig. 2.19) are retained according to Viterbi algorithm and all the other paths are discarded. In this way, lots of unwanted paths could be eliminated and an exponential searching problem is transformed into a linear one.

2.4   Summary

This chapter gives a vivid idea of an end to end ASR system. In order to construct an ASR system, it is required to know the human speech production system. The human articulatory system is modeled, and each articulatory organ is considered as a variable. The various combinations of the values of the variables produce different phones. A brief history of the evolution of the ASR system is given, and it helps to understand the chronological progress of the same. An ASR system is composed of several components. The input to the ASR system is some audio features. These features are extracted in the acoustic analysis stage. Acoustic features are then mapped to corresponding phonemes. Hidden Markov model was one of the most popular mod-

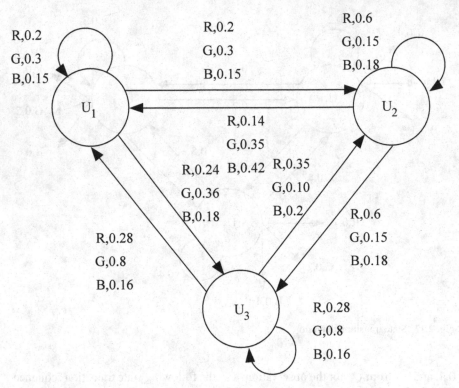

Fig. 2.18 State transition diagram absorbing the RGB probabilities

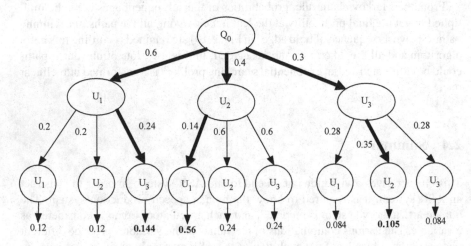

Fig. 2.19 Transition diagram for the observation "ЄR"

els. This is a statistical model to find the mapping to phonemes from feature vectors. Speech recognition has made significant progress with the introduction of neural network and deep recurrent neural network. Feedback loops in the RNN increase the speech recognition prediction significantly. The recognized phonemes are mapped to the corresponding words in the pronunciation model. Due to difference in accents, this prediction might become ambiguous. Articulatory knowledge is used here to minimize the error in prediction. Language model, on the other hand, organizes the word into meaningful sequences to form a sentence. The output of all the components of the ASR system is fed into a central decoder which is responsible to find the most accurate sentence from the large pool of sentences. This search graph is huge and there is an exponential time complexity to find the accurate hidden graph. Most of the paths in the search graph are also unwanted. Because of this, there is a need for search graph optimization. Viterbi algorithm is introduced at the end, which is an efficient algorithm for search graph optimization.

References

1. Kamm, C., Walker, M., & Rabiner, L. (1997). The role of speech processing in human–computer intelligent communication. *Speech Communication, 23*(4), 263–278.
2. Retrieved July 08, 2018, from https://www.sciencedirect.com/topics/neuroscience/speech-processing.
3. Dey, N., & Ashour, A. S. (2018). Challenges and future perspectives in speech-sources direction of arrival estimation and localization. In *Direction of arrival estimation and localization of multi-speech sources* (pp. 49–52). Cham: Springer.
4. Dey, N., & Ashour, A. S. (2018). *Direction of arrival estimation and localization of multi-speech sources.* Springer International Publishing.
5. Dey, N., & Ashour, A. S. (2018). Applied examples and applications of localization and tracking problem of multiple speech sources. In *Direction of arrival estimation and localization of multi-speech sources* (pp. 35–48). Cham: Springer.
6. Dey, N., & Ashour, A. S. (2018). Microphone array principles. In *Direction of arrival estimation and localization of multi-speech sources* (pp. 5–22). Cham: Springer.
7. Kamal, M. S., Chowdhury, L., Khan, M. I., Ashour, A. S., Tavares, J. M. R., & Dey, N. (2017). Hidden Markov model and Chapman Kolmogrov for protein structures prediction from images. *Computational Biology and Chemistry, 68,* 231–244.
8. Mahendru, H. C. (2014). Quick review of human speech production mechanism. *International Journal of Engineering Research and Development, 9*(10), 48–54.
9. Shirodkar, N. S. (2016). Konkani Speech to Text Recognition using Hidden MARKOV Model Toolit (Masters dissertation). Retrieved July 08, 2018, from https://www.kom.aau.dk/group/04gr742/pdf/speech_production.pdf.
10. Retrieved July 08, 2018, from https://www.youtube.com/watch?v=Xjzm7S__kBU.
11. Sood, S., & Krishnamurthy, A. (2004, October). A robust on-the-fly pitch (OTFP) estimation algorithm. In *Proceedings of the 12th Annual ACM International Conference on Multimedia* (pp. 280–283). ACM.
12. De Cheveigné, A., & Kawahara, H. (2002). YIN, a fundamental frequency estimator for speech and music. *The Journal of the Acoustical Society of America, 111*(4), 1917–1930.
13. Chowdhury, S., Datta, A. K., & Chaudhuri, B. B. (2000). Pitch detection algorithm using state phase analysis. *J Acoust Soc India, 28*(1–4), 247–250.

14. Yu, Y. (2012, March). Research on speech recognition technology and its application. In *2012 International Conference on Computer Science and Electronics Engineering (ICCSEE)*, (Vol. 1, pp. 306–309). IEEE.
15. Retrieved July 20, 2018, from https://www.youtube.com/watch?v=q67z7PTGRi8&t=4294s.
16. Dey, N., Ashour, A. S., Mohamed, W. S., & Nguyen, N. G. (2019). Acoustic wave technology. In *Acoustic sensors for biomedical applications* (pp. 21–31). Cham: Springer.
17. Dey, N., Ashour, A. S., Mohamed, W. S., & Nguyen, N. G. (2019). Acoustic sensors in biomedical applications. In *Acoustic sensors for biomedical applications* (pp. 43–47). Cham: Springer.
18. Khiatani, D., & Ghose, U. (2017, October). Weather forecasting using hidden Markov model. In *2017 International Conference on Computing and Communication Technologies for Smart Nation (IC3TSN)*, (pp. 220–225). IEEE.
19. Tokuda, K., Nankaku, Y., Toda, T., Zen, H., Yamagishi, J., & Oura, K. (2013). Speech synthesis based on hidden Markov models. *Proceedings of the IEEE, 101*(5), 1234–1252.
20. Retrieved July 20, 2018, from https://www.youtube.com/watch?v=kNloj1Qtf0Y&t=1500s.
21. Gales, M., & Young, S. (2008). The application of hidden Markov models in speech recognition. *Foundations and Trends® in Signal Processing, 1*(3), 195–304.
22. Rabiner, L. R., & Juang, B. H. (1992). Hidden Markov models for speech recognition—strengths and limitations. In Speech recognition and understanding (pp. 3–29). Heidelberg: Springer.
23. Hore, S., Bhattacharya, T., Dey, N., Hassanien, A. E., Banerjee, A., & Chaudhuri, S. B. (2016). A real time dactylology based feature extraction for selective image encryption and artificial neural network. In Image feature detectors and descriptors (pp. 203–226). Cham: Springer.
24. Samanta, S., Kundu, D., Chakraborty, S., Dey, N., Gaber, T., Hassanien, A. E., & Kim, T. H. (2015, September). Wooden Surface classification based on Haralick and the Neural Networks. In 2015 Fourth International Conference on Information Science and Industrial Applications (ISI), (pp. 33–39). IEEE.
25. Kotyk, T., Ashour, A. S., Chakraborty, S., Dey, N., & Balas, V. E. (2015). Apoptosis analysis in classification paradigm: a neural network based approach. In Healthy World Conference (pp. 17–22).
26. Agrawal, S., Singh, B., Kumar, R., & Dey, N. (2019). Machine learning for medical diagnosis: A neural network classifier optimized via the directed bee colony optimization algorithm. In U-Healthcare monitoring systems (pp. 197–215). Academic Press.
27. Wang, Y., Chen, Y., Yang, N., Zheng, L., Dey, N., Ashour, A. S., … & Shi, F. (2018). Classification of mice hepatic granuloma microscopic images based on a deep convolutional neural network. Applied Soft Computing.
28. Lan, K., Wang, D. T., Fong, S., Liu, L. S., Wong, K. K., & Dey, N. (2018). A survey of data mining and deep learning in bioinformatics. *Journal of Medical Systems, 42*(8), 139.
29. Hu, S., Liu, M., Fong, S., Song, W., Dey, N., & Wong, R. (2018). Forecasting China future MNP by deep learning. In Behavior engineering and applications (pp. 169–210). Cham: Springer.
30. Dey, N., Fong, S., Song, W., & Cho, K. (2017, August). Forecasting energy consumption from smart home sensor network by deep learning. In International Conference on Smart Trends for Information Technology and Computer Communications (pp. 255–265). Singapore: Springer.
31. Dey, N., Ashour, A. S., & Nguyen, G. N. Recent advancement in multimedia content using deep learning.
32. Rumelhart, D. E., Hinton, G. E., & Williams, R. J. (1986). Learning representations by back-propagating errors. *Nature, 323*(6088), 533.
33. Mohamed, A. R., Dahl, G. E., & Hinton, G. (2012). Acoustic modeling using deep belief networks. IEEE Transactions on Audio, Speech & Language Processing, *20*(1), 14–22.
34. Graves, A., Mohamed, A. R., & Hinton, G. (2013, May). Speech recognition with deep recurrent neural networks. In 2013 IEEE International Conference on Acoustics, Speech and Signal Processing (ICASSP) (pp. 6645–6649). IEEE.
35. Retrieved July 21, 2018, from https://medium.com/@ageitgey/machine-learning-is-fun-part-6-how-to-do-speech-recognition-with-deep-learning-28293c162f7a.

36. Browman, C. P., & Goldstein, L. (1992). Articulatory phonology: An overview. *Phonetica, 49*(3–4), 155–180.
37. Livescu, K., Jyothi, P., & Fosler-Lussier, E. (2016). Articulatory feature-based pronunciation modeling. *Computer Speech & Language, 36,* 212–232.
38. Retrieved July 22, 2018, from http://www.speech.sri.com/projects/srilm/.
39. Retrieved July 22, 2018, from https://kheafield.com/code/kenlm/.
40. Chen, S. F., & Goodman, J. (1999). An empirical study of smoothing techniques for language modeling. *Computer Speech & Language, 13*(4), 359–394.
41. Retrieved July 24, 2018, from https://www.slideshare.net/ssrdigvijay88/ngrams-smoothing.
42. Retrieved July 24, 2018, from https://www.inf.ed.ac.uk/teaching/courses/asr/2011–12/asr-search-nup4.pdf.
43. Viterbi, A. (1967). Error bounds for convolutional codes and an asymptotically optimum decoding algorithm. *IEEE Transactions on Information Theory, 13*(2), 260–269.
44. Gerber, M., Kaufmann, T., & Pfister, B. (2011, May). Extended Viterbi algorithm for optimized word HMMs. In 2011 IEEE International Conference on Acoustics, Speech and Signal Processing (ICASSP) (pp. 4932–4935). IEEE.

Chapter 3
Feature Extraction

3.1 Introduction

In order to classify any audio or speech signal, feature extraction is the prerequisite. The analog speech signal s(t) is sampled a number of times per second to be stored in some recording device or simply on a computer. These samples represent some numerical values computed repeatedly after certain time duration which is also called the sampling period (e.g., the sampling period for commercial audio CD is 1/44100 s). The recorded signal is a discrete-time signal. Therefore, except reconstructing the analog signal from the digitized version for hearing it from a speaker, these numerical values do not have that much significance. However, in digital signal processing applications, e.g., automatic speech recognition, speaker recognition and verification, speech transformation and conversion, the speech samples should be processed effectively based on which some definite algorithms will work. Therefore, the speech samples should be transformed into something that makes sense with respect to the speech generation or perception or more precisely in natural language processing. This "something" should differentiate or characterize a speech sample among other speech samples, which is called a feature. Thus, the features can be represented as numeric values that can model interesting parts of producing speech or notable changes that happen in a speech signal. The acoustic input signal is converted into a sequence of acoustic feature vectors by the feature extraction process, and thus, a good representation of the input speech signal is possible. Due to the large variability in speech signals, feature extraction is performed to reduce the same. The feature extraction process is of two types: temporal analysis technique and spectral analysis technique. The temporal analysis considers the waveform of the speech or audio signal for the analysis purpose. The spectral form of the audio signal is utilized by the spectral analysis.

In order to have any precise idea about the feature extraction process, one should have some basic ideas about different kinds of audio features. Section 3.2 describes some basic frequently used audio features required for speech classification process. One could have a vivid idea about basic audio features like pitch, timbral features,

© The Author(s), under exclusive license to Springer Nature Singapore Pte Ltd. 2019   45
S. Sen et al., *Audio Processing and Speech Recognition*, SpringerBriefs in
Computational Intelligence, https://doi.org/10.1007/978-981-13-6098-5_3

e.g., zero-crossing rate, spectral centroid, spectral roll-off and spectral flux, etc. Inharmonicity and autocorrelation are the two properties important for similarity analysis of the audio signals. These are discussed in Sects. 3.2.4 and 3.2.5, respectively. Section 3.2.7 elaborates MPEG-7 features which are one kind of standardization of audio features. After giving a brief idea about different audio features, it is followed by different important feature extraction techniques and their comparative studies in Sect. 3.3.

3.2 Basic Audio Features

An audio signal is divided into a number of frames before feature extraction. If the sampling frequency is 8000 Hz then each frame is about 256 samples, i.e., 32 ms each. In general, these frames or windows are around 10–40-ms in length. For each frame, there is a hamming window [1]

$$w(n) = 0.54 - 0.46 \cos\left[\frac{2\pi n}{(N-1)}\right]$$

where N is the total number of samples and n is the current sample. If the original signal be $x(n)$, then the sample data after adding the window is:

$$s(n) = x(n)w(n).$$

Thus, the audio signal is broken down into a number of analysis windows and for each window, one feature value is computed. Feature extraction could be done in two ways. In the first approach, all the values corresponding to features for a given analysis window are obtained to formulate the feature vectors in order to take the classification decision. In the second approach, a texture window is used to extract the long-term characteristics of the audio signal. Since the variation of the features w.r.t. time is measured, it is possible to obtain a better and relevant description of the signal.

In the subsequent subsections, some of the frequently used audio features for audio classification are briefly described.

3.2.1 Pitch

Chapter 2 describes the difference between voiced and unvoiced sounds. Vocal cord does not vibrate for unvoiced sounds and stays open. But during voiced sound, vocal cord vibrates and thus a glottal pulse is produced. A pulse consists of sinusoidal waves of the fundamental frequencies and its harmonics. Pitch is the fundamental frequency of the glottal pulse. Pitch helps to identify a specific tone and distinguish between different sounds from different types of sources, e.g., musical instruments, etc. Pitch could be analyzed both in time and frequency domain. Pitch could be analyzed in time domain by estimating it by using the peaks but the presence of formant frequencies might be a hindrance here. To avoid this, the formant frequencies are filtered out by using a low-pass filter. After filtering out the formant frequencies, zero-crossing method or any relevant technologies are used to determine the pitch. In the frequency domain also, the speech signal is passed through a low-pass filter, and then, the pitch is determined by doing the spectrum analysis.

3.2.2 Timbral Features

Distinguishing characteristics of sound that helps to identify different sounds having same pitch and loudness is defined as sound quality or timbre. Harmonic components of a sound mainly determine the timbre. Besides harmonic contents, vibrato and tremolo are the two dynamic characteristics of sound that determine the timbre. Different types of musical instruments could be also identifiable by observing the timbre. Vibrato is defined as the periodic changes in the pitch associated with a particular tone. On the other hand, the term tremolo indicates the periodic changes in the amplitude or loudness of the tone. So vibrato could be considered as the frequency modulation (FM) and tremolo as the amplitude modulation (AM) of the tone. Both Vibrato and Tremolo are present in the speech or music to some extent. Vibrato is one of the useful characteristics of the human voice and it could be used to increase the richness of the speech. If the harmonic content of a sample speech could be resynthesized accurately, then also the ear can identify the difference in timbre since the vibrato is missing in the generated speech.

In the below section some feature detection measures are briefly described.

(i) **Zero Crossings**

In time-domain signal analysis, the zero-crossing rate (ZCR) [2] is defined as the number of times the amplitude of the signal changes sign in one frame. For a single-voiced source, the fundamental frequency is estimated from zero crossings. In case of complex sounds, it is a measure of the presence of noise in the signal. Zero-crossing rate is a basic measure of the frequency content in the signal. ZCR could be represented as follows:

$$ZCR = \sum_{n=1}^{N-1} |sgn[x(n+1)] - sgn[x(n)]|/2(N-1)$$

where $x(n)$ is the discrete audio signal

$$sgn[(x(n)] = \begin{cases} 1(x(n) \geq 0) \\ -1(x(n) \leq 0) \end{cases}$$

The speech signal is less periodic than music or instrumental sounds. ZCR of music or instrumental sounds has less fluctuation or more constant than speech waveform.

(ii) Spectral Centroid

The center of gravity of the spectrum is defined as the spectral centroid [3]:

$$C_r = \frac{\sum_{k=1}^{\frac{N}{2}} f[k]|X_r[k]|}{\sum_{k=1}^{\frac{N}{2}} |X_r[k]|}$$

where $f[k]$ is the frequency of the kth bin, r is the current frame number, and $[k]$ indicates the short-time Fourier transform (STFT) of the frame.

Spectral shape is measured by the centroid. A high centroid value means a more prominent texture having high frequencies. The sharpness of sound is directly proportional to the spectrum frequency content. Higher centroid value relates to the high-frequency content and results in high sharpness. Since centroid is very effective to represent the spectral shape, it is used as an important feature for audio classification process.

(iii) **Spectral Roll-Off**

Spectral roll-off is the frequency which acts as an upper bound of the 85% of the spectrum magnitude distribution concentration. Like spectral centroid, the spectral shape could be also measured by this feature. Roll-off can be represented by the following equation:

$$\sum_{k=1}^{F} |X_r[k]| = 0.85 \sum_{k=1}^{\frac{N}{2}} |X_r[k]|$$

If F is the limiting value of k satisfying the above equation, then this is the roll-off frequency.

Spectral roll-off is a good measure of the spectral shape "skewness" [4]. It could be used to differentiate between voiced and unvoiced speech since unvoiced speech has a relatively higher energy distribution in the high-frequency range of the speech spectrum.

(iv) Spectral Flux

The magnitudes of the successive spectral distributions are normalized and the squared difference of them is known as spectral flux [4]. Spectral flux F_t could be represented by the following equation:

$$F_t = \sum_{n=1}^{N} (N_t[n] - N_{t-1}[n])^2$$

where $N_t[n]$ and N_{t-1} are the normalized magnitude of the Fourier transform at frames t and $(t-1)$.

3.2.3 Rhythmic Features

The regularity of the audio signal could be described by the rhythmic features. Since a particular pattern is followed, characteristics of the audio signal could be defined by the rhythmic features. There are two types of rhythmic features: rhythmical structure and bit strength.

For an effective classification of audio signals, information about these features should be extracted. Regularity of bits is higher in rock or pop music and this could be considered as an example of the rhythmic structure. Bit strength is also an important feature, e.g., bit strength of jazz is less than techno music.

Bit strength is represented as a histogram which is a curve represented by a function of a range of beats per minute values. Main bit and other sub bit are represented by the peaks in the histogram.

(i) **Bit Strength**

From the histogram, different statistical measures like mean, standard deviation, skewness and entropy, etc., are obtained to evaluate the measure of beat strength. These measures are computed in the Bit domain.

(ii) **Rhythmic Regularity**

If there is periodic spacing in a beat histogram, the peaks indicate high rhythmic regularity. The normalized autocorrelation function is computed from the beat histogram to measure the rhythmic regularity. The computation is done at the frame by frame basis, but histograms are measured in long-term intervals defined by the texture window.

3.2.4 Inharmonicity

The divergence of the spectral component of the audio signal from a pure harmonic signal is defined as the inharmonicity [5]. The deviation is well described by the following classical expression [6]:

$$f_n = nf_0\left(1 + n^2 B\right)^{0.5}$$

where f_n is the frequency of the nth partial, f_0 is the fundamental frequency, and B is the inharmonic coefficient.

3.2.5 Autocorrelation

When a signal is compared with a time-delayed version of itself, it is called autocorrelation. By using this feature, the periodic nature of a signal could be detected. If a signal is periodic then it will be perfectly correlated with its own shifted version. Thus, autocorrelation gives the degree of similarity between two audio signals.

Mathematically autocorrelation could be calculated as follows [7].

If the time delay is T, then the following steps are performed:

1. At a particular time t, the value of the signal is found.
2. The value of the signal at time (t + T) is found.
3. Above two calculated values are multiplied together.
4. Repeat steps 1–3 for all possible values of t.
5. The average of all these products is computed.

This process could be repeated for all possible values of T, which results in an autocorrelation, which is simply a function of time delay T.

For a continuous signal, f (t), the autocorrelation, F (T) can be calculated as

$$F(T) = \frac{1}{(t_{max} - t_{min})} \int_{t_{min}}^{t_{max}} f(t)f(t+T)dt$$

Autocorrelation can be used to know whether a signal is periodic or not. Suppose, for T = 0. R (0) = 1. Now for greater values of T, if R (T) also becomes always 1 then it could be decided that the signal is periodic.

Autocorrelation is very useful in voice recognition. One critical example might be, suppose someone catches a cold on a certain day. His voice might be a little bit different than his original voice (it could be considered as a lagging signal) in this condition. Now if autocorrelation is used to compare the two voices then a certain level of similarity could be found and it could be concluded that the two voices are the voices of the same person. This is because the autocorrelation is integration and the result is the area under the curve of the following expression: f(t)*f(t + T).

So comparing with cold voice will generate a similar pattern.

3.2.6   Other Features

This section describes some features which are based on the signal's predictability, statistical pattern, and dynamic characteristics. They are as follows:

(i)   Root Mean Square (RMS)
      It is the root mean square energy of the signal frame.
(ii) Time Envelope
      It indicates the maximum of absolute amplitude in each frame.
(iii) Low Energy Rate
      Within an audio file, the percentage of frames having RMS energy less than the mean RMS energy is known as low energy rate. This feature is computed based on a texture window rather than frame by frame basis.

(iv) Loudness
       The above features are dynamic and calculated based on physical measures like
       amplitude or energy. Measurement of loudness helps in better adaptation to the
       perception of sound by human ears. It could be defined as, "That attribute of
       auditory sensation in terms of which sounds can be ordered on a scale extending
       from quiet to loud" [8].

3.2.7 MPEG-7 Features

MPEG stands for Moving Pictures Experts Group that has defined an international
standard where a collection of techniques are given for analyzing and representing
the raw data in terms of certain features. The need is to standardize the features for
audio or speech signal classification. In the following section, some of the audio
features are discussed.

(i) **Audio Spectrum Centroid (ASC)**

Spectral Centroid is standardized in MPEG-7. A logarithmic frequency scaled with
1 kHz is introduced. It describes the center of force of log-frequency in a power
spectrum [5]. Audio spectrum centroid of the power spectrum at frame r is defined
as

$$ASC = \frac{\sum_{K=1}^{N/2} \log_2\left(\frac{f[k]}{1000}\right) S_r[k]}{\sum_{k=1}^{N/2} S_r[k]}$$

where S_r is the power spectrum of the framer.
    ASC acts as an indicator of the domination of the high and low frequencies in the
power spectrum.

(ii) **Audio Spectrum Spread (ASS)**

The distribution of the spectrum around the centroid is called the audio spectrum
spread (ASS). It is formulated as

$$ASS_r = \sqrt{\frac{\sum_{k=1}^{N/2}[\log_2\left(\frac{f[k]}{1000}\right) - ASC_r]^2 S_r[k]}{\sum_{k=1}^{N/2} S_r[k]}}$$

ASS helps to differentiate between a noise like signal and tone [9].

(iii) **Audio Spectrum Flatness (ASF)**

Audio spectrum flatness (ASF) is the deviation of the spectral forms w.r.t. a flat spectrum. It defines the audio signal's planeness property [9]. A flat spectrum represents
noisy signals or impulse-like signals. So high flatness indicates noisy signal and low

flatness indicates the presence of the harmonic component. The flatness of a band is computed by taking the ratio of the geometric and the arithmetic means of the power spectrum coefficients within that band.

(iv) **Harmonic Ratio (HR)**

In a spectrum, the proportion of harmonic content could be measured by the Harmonic Ratio (HR). The maximum value of the autocorrelation within the frame is known as Harmonic Ratio [10].

3.3 Feature Extraction Techniques

The major challenge of ASR system is the large-scale variability due to different types of speakers, their speaking rate, speech content, and different acoustic conditions. Due to this, speaker identification is a challenging task. The main component of speaker identification and speech recognition is feature extraction and analysis [11]. In an ASR system, the feature analysis component has a major role in the overall system performance. Therefore, for feature analysis and subsequent speech recognition, the first step is the feature extraction. In Sect. 3.2, a brief idea about major audio features has been given. In this section, various feature extraction techniques are summarized.

3.3.1 Linear Prediction Coding (LPC)

Linear prediction originates from the source-filter model proposed by Gunnar Fant in 1960 [12]. The proposed model is a linear model, considering glottis, and vocal tract as fully uncoupled.

According to the source-filter model, the speech signal is the output $Y_1[n]$ of an all-pole filter $\frac{1}{A_1(Z)}$ excited by the s[n] [12], where s[n] is the excitation signal.

It could be represented as follows [12]:

$$Y_1(Z) = S(Z)\frac{1}{1 - \sum_{K=1}^{P} \alpha_k z^{-k}} = S(Z)\frac{1}{A_p(Z)} \tag{3.1}$$

where $Y_1(Z)$ and $S(Z)$ are the z transforms of the speech and excitation signals, respectively, and p is the prediction order.

The all-pole filter $\frac{1}{A_p(Z)}$ is called the synthesis filter where $A_p(Z)$ is called as inverse filter.

The concept of linear predictability is inherent in Eq. 3.1. Speech signal can be expressed as below by applying the inverse z transform:

$$y_1[n] = s[n] + \sum_{k=1}^{p} \alpha_k y_1[n-k] \qquad (3.2)$$

Equation 3.2 indicates that the current speech sample could be estimated as the summation of the p previous samples added with some excitation factor. The term $s[n]$ is the error term and also written as $e[n]$.

LPC methods are frequently used in speech synthesis, speech coding, and speech or speaker recognition and speech storage purpose. It is one of the audio compression methods that model the human speech production process. LPC method efficiently estimates the speech parameters. The basic idea about the LPC procedure could be obtained from the source-filter model which states that the current speech sample could be closely approximated as a linear combination of the past samples [13], i.e.,

$$s(n) = \sum_{k=1}^{p} \alpha_k s(n-k) \, for \, some \, values \, of \, p \, and \, \alpha_k \, = \, predictor \, coefficients$$

For linear prediction, predictor coefficients (α_k) are computed "by minimizing the sum of squared differences (over a finite interval) between the actual speech samples and the linearly predicted ones" [14].

From Eq. 3.2, the excitation signal which is the source of the speech signal could be estimated as follows:

$$s[n] = y_1[n] - \sum_{k=1}^{p} \alpha_k y_1[n-k] \qquad (3.3)$$

Equation 3.3 basically represents the output of the inverse filter excited by the speech signal.

The overall algorithm of LPC analysis consists of two stages, namely, (i) analysis or encoding part and (ii) synthesis or decoding part. In the encoding stage, LPC determines the input signal and the filter coefficients from the received speech signal frames or blocks. In the decoding stage, the filter is rebuilt based on the computed coefficients.

Frame by frame LPC analysis helps in the decision-making process of voiced or unvoiced audio signal. To compute the correct pitch period a pitch-detecting algorithm is applied.

To know further, please go through the article "linear predictive coding" by Bradbury [15].

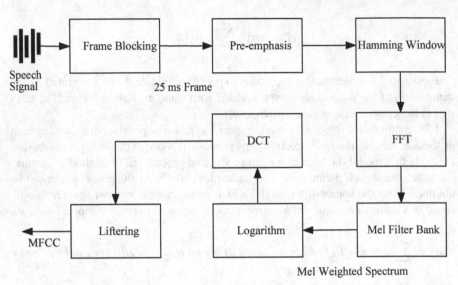

Fig. 3.1 MFCC feature extraction process [18]

3.3.2 Mel-Frequency Cepstral Coefficient (MFCC)

MFCC represents the speech signal into some parametric form [16] to easily analyze and process it. A scale of pitches which maintains equal distances between each other is called mel scale that exhibits linear feature for frequencies less than 1 kHz and logarithmic characteristics for frequencies above 1 kHz. MFCC applies mel scale for the feature extraction process because as the frequency gradually increases above 1 kHz the frequency sensitiveness of the human auditory system gradually decreases exhibiting logarithmic characteristics [17]. Figure 3.1 shows the architecture of MFCC process.

The step-by-step MFCC process as depicted in Fig. 3.1 is described below:

(i) Framing

The speech signal is not stationary if it is observed for a long period of time. Short-time analysis is done on speech signal to analyze and process it. The speech waveform is considered stable within 10–30 ms time interval. The speech signal is analyzed for frames of uniform duration (10–30 ms). In frame blocking phase the input speech signal is divided into uniform frames having N samples. The consecutive frames should be separated by a distance M where M < N [18].

(ii) Preemphasis

In the preemphasis stage, first-order differential is applied to the voiced sample to raise the energy level in the speech waveform [19]:

$$S'_n = S_n - kS_{n-1} \tag{3.4}$$

where k is the preemphasis coefficient and $0 \leq k \leq 1$.

(iii) Hamming Windowing

The next step in the MFCC process is windowing each frame. This is required because processing of the input audio signal is done on some limited number of samples. This results in discontinuities among the frames. Windowing each frame helps to reduce the discontinuities and it also makes the end of frames smooth enough so that they can accurately connect with the beginning of the next frame [19]. In the beginning and the end of each frame, a hamming window is introduced to gradually decrease the signal to zero. The following equation is applied to the voice samples $\{S_n, n = 1 \ldots \ldots N\}$ [19]:

$$S'_n = \left\{ 0.54 - 0.46 \cos \left(\frac{2\pi (n-1)}{N-1} \right) \right\} S_n \tag{3.5}$$

(iv) Fast Fourier Transform (FFT):

In this step, Fourier transform is applied to transform each frame having N samples from time domain to frequency domain. Fast Fourier transform is a better choice to do that because it is significantly fast algorithm to compute the discrete Fourier transform (DFT) [20]. On a set of N samples (S_n), the DFT could be defined by the following equation [20]:

$$S_n = \sum_{k=0}^{N-1} s_k e^{\frac{-2\pi jkn}{N}} \quad where \, n = 0, 1, 2, \ldots, N-1 \tag{3.6}$$

(v) Filter Bank Analysis:

As earlier mentioned, psychological research reveals that the human auditory system can interpret pitch according to a logarithmic scale rather than a linear scale [21]. A subjective pitch is calculated based on the Mel scale for each frequency measurement in Hertz. The following equation is used to compute a Mel corresponding to a given frequency in Hertz [21]:

$$F_{mel} = \frac{1000}{\log(2)} \cdot \left[1 + \frac{F_{hz}}{1000} \right] \tag{3.7}$$

where F_{mel} is the equivalent mel scale frequency, and F_{hz} is the normal frequency in Hertz.

(vi) Logarithm/DCT

This phase involves in calculating the mel-frequency cepstral coefficients (MFCCs) by applying discrete cosine transform (DCT) on log filter bank amplitudes a_j [22]:

$$c_i = \sqrt{\frac{2}{N}} \sum_{j=1}^{N} a_j \cdot \cos \left(\frac{\pi i}{N} (j - 0.5) \right) \tag{3.8}$$

where N is the total number of filter bank channels.

(vii) **Liftering**

Several studies [23–27] suggest that if cepstral coefficients are used as features in speech recognition, then the performance could be greatly improved by liftering the cepstral coefficients. If the ith cepstral coefficient is x_i, then the liftered coefficient is given by [28]

$$x_i' = w_i x_i$$

where w_i are the weights for i = 1, 2,…, d, and d is the dimensionality of the feature space.

If sinusoidal lifters are used, then cepstral coefficient rescaling could be realized by the following equation [25]:

$$w_i = 1 + \frac{d}{2} \sin\left(\frac{\pi i}{d}\right) \tag{3.9}$$

3.3.3 *Perceptual Linear Prediction (PLP)*

PLP was introduced by Hermansky [29]. The human speech is modeled according to the psychophysics of hearing. PLP increases the accuracy of speech recognition by discarding the irrelevant portion of the speech. The transformation of the spectral characteristics of speech to closely follow the human auditory system is the main idea of the PLP process. In Fig. 3.2, different stages of the PLP process are shown.

The first phase is the same for all the feature extraction process, i.e., framing which is already discussed in the previous section. Once the framing is done, the frames are processed through a Hamming Window and DFT is applied to the frames. Next step is the computation of the power spectrum which is obtained as follows [29]:

$$P(w) = Re(S(w))^2 + Im(S((w))^2 \tag{3.10}$$

The important stages of the PLP process are discussed below:

(i) **Critical-Band Integration (Bark Frequency Weighing)**

Eberhard Zwicker proposed bark scale [30] in 1961. In this frequency scale, equal distance indicates equal distance perceptually. This scale becomes nearly equal to a logarithmic frequency axis for frequency above 500 Hz. It represents linear characteristics below 500 Hz [2]. In this stage of PLP method, the frequency is converted to bark frequency which is a better representation of human hearing resolution. The speech signal is processed through some equally spaced trapezoidal filters in bark scale.

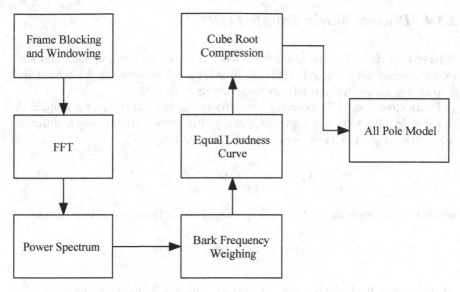

Fig. 3.2 PLP feature extraction stages

(ii) Equal Loudness Preemphasis

The sensitivity of human hearing varies for different frequencies at about 40 dB level which is represented by an equal loudness curve [31]. In a speech spectrum, the various frequency components are preemphasized by the equal loudness curve. This is generated by a filter having the specific transfer function.

(iii) Intensity Loudness Conversion (Cube-Root Amplitude Compression)

The relationship between the intensity of speech and corresponding perceived loudness is nonlinear which is also called the power law of hearing [32]. The transformed speech from the above stage is subjected to cube-root compression accordingly. The cumulative effect of equal loudness preemphasis and cube-root compression helps to reduce the variation in amplitude of the critical-band spectrum.

(iv) All-Pole Model

The final phase of the PLP method is represented by an autoregressive model which approximates the spectrum by the all-pole model with the help of the autocorrelation method. Inverse discrete Fourier transformation is done on the spectrum samples. If the all-pole model is of Nth order, then only the initial (N + 1) autocorrelation values are needed.

3.3.4   Discrete Wavelet Transform (DWT)

Wavelets are defined as the mathematical functions which break up data into various frequency components [34]. These individual components are further studied considering the particular resolution matching its scale [33].

Fourier transform (FT) converts a signal from time domain to frequency domain. A signal is decomposed into complex exponential functions corresponding to different frequencies. Below two equations define FT, respectively [34].

$$X_F(f) = \int_{-\infty}^{\infty} x_F(t).e^{-2j\pi ft} dt \tag{3.11}$$

where t represents time and f is for frequency, $x_F(t)$ is the signal in time domain

$$x_F(t) = \int_{-\infty}^{\infty} X_F(f).e^{2j\pi ft} df \tag{3.12}$$

where f stands for frequency and $x_F(f)$ is the signal in frequency domain.

Equation 3.11 represents the Fourier Transform whereas Eq. 3.12 stands for Inverse Fourier transform.

From Eq. 3.11, it is evident that the integral corresponds to all possible time instances from minus infinity to plus infinity. The location of the frequency in the time axis does not have an effect on the integral result, i.e., the presence of a certain frequency f at time t_1 or at time t_2 does not have an effect on the overall result of the integration. For this particular reason Fourier transform is not a suitable method to analyze the nonstationary signals, i.e., those frequencies vary with time. Fourier Transform only makes sense to the stationary signal where the same frequency is prevalent for the entire time duration. The Fourier Transform is useful to indicate whether a certain frequency component exists or not.

To resolve the above problem, short-time Fourier transform (STFT) is introduced to include the time dimension in the frequency plot generated by the Fourier transform. The basic essence of STFT is to consider some part of the nonstationary signal as stationary [34]. The total signal is divided into small enough parts or segments where the nonstationary signal could be considered as stationary. A window function (w) is selected for this purpose [34]. The width of this window should be carefully selected so that the signal in this portion could be considered as stationary. The window function is multiplied with the signal and FT of the product is taken. The signal is then shifted to a new position (for some t_t time unit) and multiplied by the window function. After that again FT is taken from the new product. These procedures are done repetitively until the end of the signal is reached.

The following equation computes the STFT of a nonstationary signal [34]:

$$STFT_x^w(t, f) = \int [x(t).w^*(t - t')].e^{-j2\pi ft} dt \tag{3.13}$$

where $x(t)$ is the nonstationary signal, $w(t)$ is the window function, and $*$ is the complex conjugate.

In this way, for each and every time interval t and f, new STFT coefficient is computed. By applying STFT, time–frequency plot of an audio signal is obtained.

Through STFT, the time–frequency plot could be achieved, but STFT has the following problem: It is related to the Heisenberg Uncertainty Principle which is associated with the momentum and position of the particle in motion. The same could be applied to the time–frequency plot of a signal. According to this principle, it is not possible to know the exact time–frequency information of a signal, only the time interval in which a band of frequencies exists could be obtained. The problem with the STFT originates from the window function width. In FT existence of the exact frequency is already known, there is no resolution problem in the frequency domain. Similarly, in the time domain, the exact value of a signal is known at each and every time instance, and hence, also there is no resolution problem here. In STFT, there is a window of finite length covering only one portion of the signal. Exact frequency component in this window could not be obtained; one can only know a band of frequencies.

Speech features extracted via Fourier transforms (FTs), STFTs, MFCC, or LPC techniques are applicable for certain kinds of automatic speech recognition systems where the speech signal is considered as stationary within a given time frame. They, in most of the cases, are not suitable for representing real-time speech/voice because they are unable to do an accurate analysis of localized events [35]. MFCC is the most popular technique compared to other techniques like LPC, PLP, etc., for feature extraction. As discussed in Sect. 3.3.2 MFCC is calculated by computing the discrete cosine transform (DCT) of mel-scaled log filter bank energies. This process has some drawbacks [40]. In DCT all frequency bands are covered by the basis vector, hence all MFCC are affected by the corruption of a single frequency band. So if a transform could be found in which the corruption of a frequency band only affects a nominal number of coefficients, then the effect of noise could be significantly decreased. Sometimes it may happen that a speech frame may contain information of the two adjacent phonemes. It may be possible among these two adjacent phonemes, one is voiced and another one is unvoiced. In this case the problem could be the voiced sound may dominate the low-frequency spectrum whereas the unvoiced sound may dominate the high-frequency spectrum. But in case of MFCC, a speech frame can transmit information of one phoneme at a time. If the frequency band could be divided into sub-bands and each sub-band is processed in isolation, then this asynchrony could be resolved.

As mentioned above, for various types of problems in signal processing domain, wavelet analysis [36] is proven to be an efficient one. In some studies, it is found that the wavelet coefficients could be augmented with the original feature space and a significantly less number of robust feature set is generated for the classification purpose [37–39]. In wavelet transformation, short windows are used to measure the high-frequency components of the signal and long windows are used to compute the low-frequency component of the signal. Fourier transform and short-time Fourier transform are different from the wavelet transform specifically for this property. Z

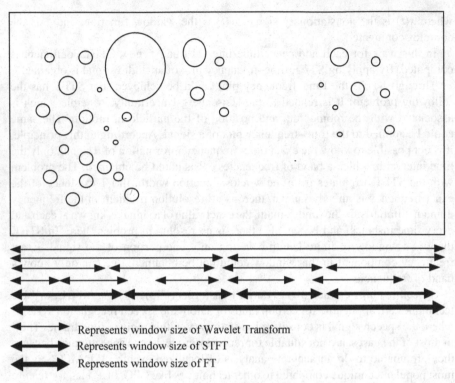

Represents window size of Wavelet Transform
Represents window size of STFT
Represents window size of FT

Fig. 3.3 Illustrative example of the difference between FT, STFT, and WT by a field with different sized rocks [40]

Tufekci and J.N. Gowdy in their work [40] clearly explained this with the example of different sized rocks in a field (Fig. 3.3).

The problem is to find the information about the size and location of the rocks. If Fourier transform is used for each size, the corresponding number of rocks could be calculated but no information about the location of the rocks could be found. If short-time Fourier transform is used, a rough estimation of the location of the rocks could be obtained along with the number of rocks for each size. Sometimes, it might happen that the rock size is bigger than the window size, and then some information about the rock would be lost. The wavelet transform is superior to Fourier transform or STFT because it uses a small window for small rocks and large window for comparatively large rocks. So if wavelet transform is used, then problems related to FT and STFT could be resolved.

Discrete wavelet transform (DWT) [41, 42] is one of the most effective mathematical transformations that provide time as well as frequency information of the input waveform. DWT is one kind of wavelet transform where discrete samples of a wavelet are obtained [43]. The input speech signal $(S(n))$ is filtered by a high-pass filter (H (z)) and by a low-pass filter (L (z)). This filtered outcome is down sampled by 2. The maximum portion of the speech energy is concentrated in the low-frequency

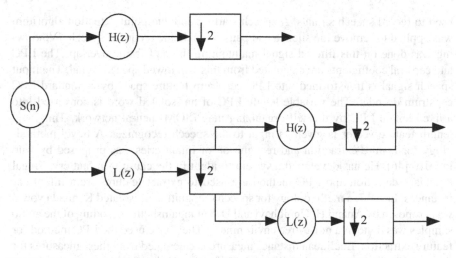

Fig. 3.4 Three stages of the analysis tree for discrete wavelet transform [43]

band and in this part, again the low-pass and high-pass filters are applied to get the sub-bands having more noise-free speech information. This procedure continues until the expected decomposition level is achieved. The mentioned DWT process is shown in Fig. 3.4. DWT exhibits good time resolution and poor frequency resolution in high frequencies and at low frequencies, DWT gives a good frequency resolution and poor time resolution. DWT could be defined by the equation below:

$$W(i, j) = \sum_i \sum_j s(j) 2^{-\frac{i}{2}} \psi \left(2^{-i} n - j \right) \tag{3.14}$$

where $\psi(t)$ is called the mother wavelet which is a time function having finite energy and rapid decay, and i is the wavelet scale and j is the translation factor of the wavelet.

There are several audio or speech feature extraction methodologies. Some of them are discussed in the above section. Several researches have been done on feature extraction techniques to improve the accuracy of automatic speech recognition (ASR) system and audio classification techniques. Some of the notable works are discussed below.

In 2003, a novel methodology for developing an improved Linear Predictive Method for ASR system was proposed by Jiang Hai et al. [44]. The discriminant analysis (LDA) was used to increase the distinguishable characteristics of the LPC coefficients and to increase the accuracy of the speech classifier. LDA was used to compress the frame vector space comprising of static and dynamic LPC coefficients. Thus, the dimension was reduced, and the resulting feature space was applied in the speech recognition system. A speech recognition system for Bengali language using LPC and Artificial Neural Network (ANN) was proposed by Anup Kumar Paul et al. in 2009 [45]. The recognition system was divided into two major parts: signal processing part and pattern recognition system. An audio wave recorder was used to record speech signals. A speech starting and end point detection algorithm was applied to remove the silence and pauses inside the recorded speech. Windowing was done on this filtered signal maintaining the half frame overlap. The LPC and cepstral coefficients are calculated from this windowed speech signal. The input speech signal is transformed into LPC cepstrum feature space by a standard LPC cepstrum decoder. The variable length LPC of an isolated word is converted into a fixed length LPC by the self-organizing map (SOM) neural network. This fixed length feature vector is given as input to the speech recognizer. A novel methodology for feature extraction and recognition of infant cries was proposed by Kuo in 2010 [46]. He has developed a system to identify the cause of infant cry. Singal boundary detection and LPC method are used to extract features from infant cry instances. In 2017, a methodology for speech recognition of isolated Kannada vowel was proposed by Anand H. Unnibhavi and D. S. Jangamshetti. Recoding of the audio samples was done in a noise-free environment. They have used the LPC method for feature extraction. Euclidean distance measure is calculated from these measures for the classification purpose. A new approach for developing feature extraction based on MFCC was proposed by Hossan et al. [47]. Discrete Cosine Transform was used here in addition to the MFCC. Three types of feature extraction techniques: MFCC, Delta-Delta MFCC, and distributed DCT were proposed to do speaker verification tests. An improved version of the MFCC was proposed by Wanli and Guoxin in 2013 [48]. They have used an experimental database comprising 15 male and 15 female speakers in a soundproof room. MFCC was used to parameterize the sample speech frames. Each frame of speech was represented by a 30-dimensional feature vector which consists of 15 MFCC with first differentials appended. In 2015, a modified version of the MFCC feature extraction technique was proposed by Sharma and Ali [49]. A different filtering approach was used here, resulting in a complementary feature set based on MFCC. On the nonlinear Mel scale, the filter banks were positioned linearly to generate a new feature set. One of the popular speech coders in communication systems is code-excited linear predictive coder (CELP). In 2009, a PLP-based CELP coder was proposed by Najafi and Marvi [50]. They have determined the required parameters for the formant synthesis filter with the help of PLP algorithm. Thereafter these parameters are applied in CELP to improve its performance. In 2012 different features extraction techniques were analyzed by Josef Chaloupka et al. to improve the system for automatic lecture transcription [51]. They have extracted features based on MFCC and PLP. A database was created containing several audio recordings of the archived audio lectures. The speech recognizer was tested with different speech features and it was observed that a higher recognition accuracy could be achieved by using PLP-based audio features. A power-law adjusted linear prediction methodology for feature extraction was proposed by Saeidi et al. in 2016 [52]. The proposed PLP model was applied to analyze the loud or shouted speech and normal loudness speech to do efficient speaker recognition. An adjusted LP is used where the speech power spectrum was adjusted by power nonlinearity. In recent years, discrete wavelet transform (DWT) has gained much popularity as an efficient feature extraction method. In 2010, DWT-based reverberation feature extraction from music recordings was proposed by Gang et al. [53]. First, the audio signal is transformed into a two-dimensional time–frequency representation by performing short-time Fourier transform. In this representation, the reverberation features seems to be a blurred spectral features. The STFT image was transformed to wavelet domain by applying image analysis method. Thereafter accurate edge detection and characterization could be performed. In 2012, a new method for feature extraction to recognize the isolated word by using DWT and LPC method was proposed by Nehe and Holambe [54]. Their proposed model was based on both wavelet decomposition and LPC coefficients. Instead of directly extracting LPC coefficients they have extracted the coefficients from the speech frames decomposed by applying DWT. A better speech recognition rate and reduced feature vector dimension could be achieved by the proposed model. A model for feature extraction and classification of the Indonesian syllables was proposed by Kristomo et al. in 2016 [55]. DWT and statistical methods were combined to recognize the Indonesian syllables. They have used three different mother wavelet transforms in combination with the statistical methods to extract the features.

3.4 Summary

Classification of the audio or speech signal is based on some audio signal parameters which are the audio features. This chapter vividly describes different audio features and various feature extraction techniques. Section 3.2 begins by explaining very basic concepts like framing and gradually digs deep into the various audio features. Pitch is responsible for identifying sounds from different audio sources. It could be analyzed both in time and frequency domain. Timbral features are the features which are useful to distinguish different sounds having the same pitch and tone. Zero crossings, spectral centroid, spectral roll-off, and spectral flux are some of the important timbral features. Rhythmic features are based on the regularity of the audio signal. One can distinguish, e.g., sounds from different instruments by studying their rhythmic features. Bit strength and rhythmic regularity are the two important rhythmic features. Section 3.2.4 describes the feature inharmonicity, which is a measure of the deviation of an audio signal from a purely harmonic one. Section 3.2.5 gives a brief idea about autocorrelation, which is a very important similarity measure between two audio signals. There was a need to standardize the features for maintaining uniformity in the feature extraction techniques. Section 3.2.7 discusses the different features of MPEG-7 standard. In Sect. 3.3, different feature extraction techniques are discussed. Section 3.3.1 describes linear predictive coding which basically originates from the source-filter model. The main aim of the LPC model is to estimate the current speech sample by analyzing the previous speech samples. Section 3.3.2 gives a brief idea about mel-frequency cepstral coefficient (MFCC). MFCC is based on mel scale which replicates the frequency sensitiveness of the human auditory system. The step-by-step procedure of MFCC is discussed here. In Sect. 3.3.3, a vivid description of perceptual linear prediction (PLP) is given. PLP is the extension of the linear prediction where the human perception of speech is considered. All these methods are only applicable for analyzing the stationary signal. But in real time, in many cases, one might have to work with the nonstationary speech signals. Section 3.3.4, first elaborates the limitation of the Fourier transform and short-time Fourier transform for analyzing a nonstationary audio signal. After that, it also highlights the problem regarding MFCC in analyzing nonstationary audio signal. At the end of this chapter, the concept of wavelets is introduced, and the discrete wavelet transform is briefly explained to analyze real-time-varying speech signal.

References

1. De Poli, G., & Mion, L. (2006). *From audio to content. Livro não publicado.* Padova: Dipartimento di Ingegneria Dell'Informazione-Università degli Studi di Padova.
2. Song, Y., Wang, W. H., & Guo, F. J. (2009). Feature extraction and classification for audio information in news video. In *2009 International Conference on Wavelet Analysis and Pattern Recognition, ICWAPR 2009* (pp. 43–46). IEEE.
3. Burred, J. J., & Lerch, A. (2004). Hierarchical automatic audio signal classification. *Journal of the Audio Engineering Society, 52*(7/8), 724–739.
4. Tzanetakis, G., & Cook, P. (2002). Musical genre classification of audio signals. *IEEE Transactions on Speech and Audio Processing, 10*(5), 293–302.
5. Galembo, A., & Askenfelt, A. (1994). Measuring inharmonicity through pitch extraction. *Journal STL-QPSR, 35*(1), 135–144.
6. Fletcher, H. (1964). Normal vibration frequencies of a stiff piano string. *The Journal of the Acoustical Society of America, 36*(1), 203–209.
7. Retrieved September 09, 2018, from https://pages.mtu.edu/~suits/autocorrelation.html.
8. American National Standards Institute. (1973). American national psychoacoustical terminology S3. 20.
9. Lazaro, A., Sarno, R., Andre, R. J., & Mahardika, M. N. (2017). Music tempo classification using audio spectrum centroid, audio spectrum flatness, and audio spectrum spread based on MPEG-7 audio features. In *2017 3rd International Conference on Science in Information Technology (ICSITech)* (pp. 41–46). IEEE.
10. Burred, J. J., & Lerch, A. (2004). Hierarchical automatic audio signal classification. *Journal of the Audio Engineering Society, 52*(7/8), 724–739.
11. Chauhan, P. M., & Desai, N. P. (2014). Mel frequency cepstral coefficients (mfcc) based speaker identification in noisy environment using wiener filter. In *2014 International Conference on Green Computing Communication and Electrical Engineering (ICGCCEE)* (pp. 1–5). IEEE.
12. Lindblom, B., Sundberg, J., Branderud, P., Djamshidpey, H., & Granqvist, S. (2010). The Gunnar Fant legacy in the study of vocal acoustics. In *10ème Congrès Français d'Acoustique*.
13. Retrieved September 13, 2018, from https://www.yumpu.com/en/document/view/18555951/17-linear-prediction-of-speech.
14. Retrieved September 12, 2018, from https://www.ece.ucsb.edu/Faculty/Rabiner/ece259/speech%20course.html.
15. Bradbury, J. (2000). *Linear predictive coding.* Hill: Mc G.
16. Kumar, C. S., & Rao, P. M. (2011). Design of an automatic speaker recognition system using MFCC, vector quantization and LBG algorithm. *International Journal on Computer Science and Engineering, 3*(8), 2942.
17. Shrawankar, U., & Thakare, V. M. (2013). Techniques for feature extraction in speech recognition system: A comparative study. arXiv preprint arXiv:1305.1145.
18. Benba, A., Jilbab, A., & Hammouch, A. (2014). Voice analysis for detecting persons with Parkinson's disease using MFCC and VQ. In *The 2014 International Conference on Circuits, Systems and Signal Processing* (pp. 23–25).

19. Young, S., et al. (2006). *The HTK book (v3. 4)*. Cambridge University.
20. Brigham, E. O., & Morrow, R. E. (1967). The fast Fourier transform. *IEEE spectrum, 4*(12), 63–70.
21. Retrieved September 16, 2018, from kom.aau.dk/group/04gr742/pdf/MFCC_worksheet.pdf.
22. Wanli, Z., & Guoxin, L. (2013). The research of feature extraction based on MFCC for speaker recognition. In *2013 3rd International Conference on Computer Science and Network Technology (ICCSNT)* (pp. 1074–1077). IEEE.
23. Paliwal, K. K. (1982). On the performance of the quefrency-weighted cepstral coefficients in vowel recognition. *Speech Communication, 1*(2), 151–154.
24. Tohkura, Y. (1987). A weighted cepstral distance measure for speech recognition. *IEEE Transactions on Acoustics, Speech, and Signal Processing, 35*(10), 1414–1422.
25. Juang, B. H., Rabiner, L., & Wilpon, J. G. (1986). On the use of bandpass liftering in speech recognition. In *IEEE International Conference on ICASSP'86 Acoustics, Speech, and Signal Processing* (Vol. 11, pp. 765–768). IEEE.
26. Itakura, F., & Umezaki, T. (1987). Distance measure for speech recognition based on the smoothed group delay spectrum. In *IEEE International Conference on ICASSP'87 Acoustics, Speech, and Signal Processing* (Vol. 12, pp. 1257–1260). IEEE.
27. Hanson, B., & Wakita, H. (1987). Spectral slope distance measures with linear prediction analysis for word recognition in noise. *IEEE Transactions on Acoustics, Speech, and Signal Processing, 35*(7), 968–973.
28. Paliwal, K. K. (1999). Decorrelated and liftered filter-bank energies for robust speech recognition. In *Sixth European Conference on Speech Communication and Technology*.
29. Hermansky, H. (1990). Perceptual linear predictive (PLP) analysis of speech. *The Journal of the Acoustical Society of America, 87*(4), 1738–1752.
30. Zwicker, E. (1961). Subdivision of the audible frequency range into critical bands (Frequenzgruppen). *The Journal of the Acoustical Society of America, 33*(2), 248–248.
31. Hermes, D. J. *Sound Perception: The Science of Sound Design*.
32. Stevens, S. S. (1957). On the psychophysical law. *Psychological Review, 64*(3), 153.
33. Graps, A. (1995). An introduction to wavelets. *IEEE Computational Science and Engineering, 2*(2), 50–61.
34. Polikar, R. (1996). Fundamental concepts & an overview of the wavelet theory. In *The Wavelet Tutorial Part I. Rowan University, College of Engineering Web Servers* (vol. 15).
35. Avci, E., & Akpolat, Z. H. (2006). Speech recognition using a wavelet packet adaptive network based fuzzy inference system. *Expert Systems with Applications, 31*(3), 495–503.
36. Siafarikas, M., Ganchev, T., & Fakotakis, N. (2004). Wavelet packet based speaker verification. In *ODYSSEY04-The Speaker and Language Recognition Workshop*.
37. Buckheit, J. B., & Donoho, D. L. (1995). Wavelab and reproducible research. In *Wavelets and statistics* (pp. 55–81). New York: Springer.
38. Wesfreid, E., & Wickerhauser, M. V. (1993). Adapted local trigonometric transforms and speech processing. *IEEE Transactions on Signal Processing, 41*(12), 3596–3600.
39. Visser, E., Otsuka, M., & Lee, T. W. (2003). A spatio-temporal speech enhancement scheme for robust speech recognition in noisy environments. *Speech Communication, 41*(2–3), 393–407.
40. Tufekci, Z., & Gowdy, J. N. (2000). Feature extraction using discrete wavelet transform for speech recognition. In *Proceedings of the IEEE Southeastcon 2000* (pp. 116–123). IEEE.
41. El-Attar, A., Ashour, A. S., Dey, N., Abdelkader, H., Abd El-Naby, M. M., & Sherratt, R. S. (2018). Discrete wavelet transform-based freezing of gait detection in Parkinson's disease. *Journal of Experimental & Theoretical Artificial Intelligence*, 1–17.
42. Mukhopadhyay, S., Biswas, S., Roy, A. B., & Dey, N. (2012). Wavelet based QRS complex detection of ECG signal. arXiv preprint arXiv:1209.1563.
43. Rady, E. R., Yahia, A. H., El-Sayed, A., & El-Borey, H. Speech recognition system based on wavelet transform and artificial neural network.
44. Zbancioc, M., & Costin, M. (2003). Using neural networks and LPCC to improve speech recognition. In *2003 International Symposium on Signals, Circuits and Systems, SCS 2003* (Vol. 2, pp. 445–448). IEEE.

45. Paul, A. K., Das, D., & Kamal, M. M. (2009). Bangla speech recognition system using LPC and ANN. In *Seventh International Conference on Advances in Pattern Recognition, 2009. ICAPR'09* (pp. 171–174). IEEE.
46. Kuo, K. (2010). Feature extraction and recognition of infant cries. In *2010 IEEE International Conference on Electro/Information Technology (EIT)* (pp. 1–5). IEEE.
47. Hossan, M. A., Memon, S., & Gregory, M. A. (2010). A novel approach for MFCC feature extraction. In *2010 4th International Conference on Signal Processing and Communication Systems (ICSPCS)* (pp. 1–5). IEEE.
48. Wanli, Z., & Guoxin, L. (2013). The research of feature extraction based on MFCC for speaker recognition. In *2013 3rd International Conference on Computer Science and Network Technology (ICCSNT)* (pp. 1074–1077). IEEE.
49. Sharma, D., & Ali, I. (2015). A modified MFCC feature extraction technique for robust speaker recognition. In *2015 International Conference on Advances in Computing, Communications and Informatics (ICACCI)* (pp. 1052–1057). IEEE.
50. Najafi, J., & Marvi, H. (2009). PLP based CELP speech coder. In *2009 Second International Conference on Computer and Electrical Engineering, ICCEE'09* (Vol. 1, pp. 263–267). IEEE.
51. Chaloupka, J., Červa, P., Silovský, J., Žd'ánský, J., & Nouza, J. (2012). Modification of the speech feature extraction module for the improvement of the system for automatic lectures transcription. In *ELMAR, 2012 Proceedings* (pp. 223–226). IEEE.
52. Saeidi, R., Alku, P., & Bäckström, T. (2016). Feature extraction using power-law adjusted linear prediction with application to speaker recognition under severe vocal effort mismatch. *IEEE/ACM Transactions on Audio, Speech and Language Processing (TASLP), 24*(1), 42–53.
53. Gang, R., Bocko, M. F., & Headlam, D. (2010). Reverberation features identification from music recordings using the discrete wavelet transform. In *2010 IEEE International Conference on Acoustics Speech and Signal Processing (ICASSP)* (pp. 161–164). IEEE.
54. Nehe, N. S., & Holambe, R. S. (2012). DWT and LPC based feature extraction methods for isolated word recognition. *EURASIP Journal on Audio, Speech, and Music Processing, 2012*(1), 7.
55. Kristomo, D., Hidayat, R., & Soesanti, I. (2016). Feature extraction and classification of the Indonesian syllables using Discrete Wavelet Transform and statistical features. In *International Conference on Science and Technology-Computer (ICST)* (pp. 88–92). IEEE.

Chapter 4
Audio Classification

4.1 Introduction

Classification falls under supervised learning. Supervised learning is a learning process from a given dataset or training dataset where both input and mapping output data are provided. The decision rules are designed by observing the training dataset to determine the category or class for future decision-making. Classification is the process of assigning an individual item or dataset to one of the number of existing categories or classes depending on the characteristics or features of the input data. The classification process consists of two steps: constructing the classifier model and using the classifier model for classification. Constructing the classification model goes through a learning phase. Classification algorithm analyzes the training dataset and builds the classifier from the given training set consisting of database tuples and their corresponding class labels. Once the classifier is built from the training dataset, test data is used to estimate the accuracy of the model. If the accuracy is above some predetermined threshold then the classification rules are acceptable. The classification rules could be applicable to the new dataset if the accuracy is acceptable. One of the most growing and popular research and application area of speech recognition is the classification phase. This chapter summarizes some of the state-of-the-art classification techniques and their application in automatic speech recognition system. In this chapter, various classification techniques and corresponding parameter estimation methods are discussed. Along with it, research and application areas of each classification method are discussed.

Section 4.2 discusses some traditional classification techniques and related research works in speech classification. The basic and most simple classification technique is k-nearest-neighbor (kNN) algorithm which is discussed in Sect. 4.2.1. Naïve Bayes classification is elaborated in Sect. 4.2.2 which is used for high-dimensional inputs. Section 4.2.3 illustrates the decision tree which is a prediction-based analytical and classification tool. A support vector machine (SVM) acts as a discriminative classifier by categorizing the training examples with the help of a separating hyperplane. Section 4.2.4 gives a vivid description of the SVM classification technique.

After illustrating the classification techniques, related research works on speech classification is discussed to relate the application of the traditional classification techniques with speech classification. With the advancement of neural network, speech recognition methodologies have been improved drastically. Several researches have been done to build neural network classification model; some of these are highlighted in Sect. 4.3. Neural network models are used as an alternative to traditional classifier and in some cases, a combination of statistical models like hidden Markov model (HMM) and neural network model is applied in order to perform automatic speech recognition (ASR). The inherent dynamic characteristics of speech attract the research on application of deep neural network (DNN) in ASR. Some of the related state-of-the-art researches are discussed in Sect. 4.4. This section establishes the fact that DNN could be used as a suitable alternative to traditional speech classification models.

4.2   Classification Strategies

Several pioneering classification strategies are discussed in the below section.

4.2.1   k-Nearest Neighbors (k-NN)

The basic and most simple classification technique is K-nearest neighbor (k-NN) algorithm [1]. It is applicable to the systems where the knowledge about the distribution of data is poor or little. This methodology was developed to perform discriminative analysis of data when no parametric estimation of probability densities is available or the same is difficult to estimate.

In an unpublished US Air Force School of Aviation Medicine report, Fix and Hodges introduced a nonparametric-based pattern classification algorithm which is the basic building block of the k-NN algorithm [2, 3]. In 1967 Cover and Hart proposed some formal properties of the k-NN rule and showed that error probability of k-NN is bounded above twice the Bayes probability of error [4]. Once the basic properties of the k-NN was decided; it has been refined in the subsequent researches and improved over the time. New issues are identified and accordingly, these are included in researches. Some of these concepts are mentioned below:

- k-NN including new rejection approaches [5],
- refinements with respect to Bayes error rate [6],
- distance weighted approaches [7, 8],
- soft computing [9] methods, and
- fuzzy methods [10].

k-NN is a nonparametric, lazy, and instance-based supervised machine learning algorithm used for classification. The term nonparametric means that, in k-NN, no assumption is made regarding the data under consideration. In many scenarios, real-time data does not obey or maintain the predefined theoretical assumption, and a significant deviation is seen. This results in an erroneous result. k-NN considers current data for classification purpose. Since raw training data or instance is used for the prediction process, it is also called instance-based algorithm.

k-NN procedure does not need any previous training. Instead of learning from the previous dataset to do some generalization, k-NN stores the entire dataset. Since no learning process is there in k-NN, all the necessary work is done at the time of prediction. For this reason, k-NN is also called a lazy process.

In k-NN, the data exists in a feature space or in a metric space. There might be scalar data or multidimensional data. Since computation has to be done in a feature space, there is a necessity of the notion of a distance measurement. The training dataset includes set of vectors and associated class labels. k-NN directly predicts from the training dataset. In order to predict the class of a new instance (p) exhaustive search is done through the entire training dataset to find the k most nearest neighbors of the new instance. Distance measures are used to determine which neighbors are more close or similar to the new item or instance under consideration. The most popular and widely used distance measure is Euclidean distance.

The square root of the summation of the squared distances of the new points and the existing points is called the Euclidean distance. Euclidean distance (d) between two points $p_i(x_1, y_1)$ and $p_j(x_2, y_2)$ in a two-dimensional plane is defined as

$$d = \sqrt{(x_1 - x_2)^2 + (y_1 - y_2)^2} \qquad (4.1)$$

The value of k can be determined by tuning the algorithm. It is a good practice to try with different values of k and choose that value of k which works best for the particular algorithm. Choosing the appropriate value of k is one of the vital decision-making since it refers to the number of nearest neighbors to be included in the classification process. k-NN method is based on feature similarity. Choosing the appropriate value of k is called parameter tuning, and it is most needed for better accuracy. Instead of tuning, there is one hard-and-fast rule to choose k (however, it is not mandatory). If the total number of data points be n, then $k = \sqrt{n}$. Generally, odd value of k is selected to remove the ambiguity between two classes of data.

k-NN process is explained by the following example.

Suppose the training data consists of n people having different heights and weights. The training dataset contains two variables: height and weight based on which a person could be identified as healthy, underweight or obese. A sample dataset is given in Table 4.1.

k-NN procedure is applied to the above training dataset to find out the class of a new person having height and weight as 169 and 69, respectively. Euclidean distance between the unknown data point from the all data points in the training dataset should be calculated.

Table 4.1 Training dataset

Person	Weight (x_n)	Height (y_n)	Class/Label
1	53	170	Underweight
2	65	185	Healthy
3	67	173	Healthy
4	50	167	Underweight
5	66	175	Healthy
6	54	173	Underweight
7	81	168	Obese
8	85	170	Obese

Table 4.2 Euclidean distance calculation on sample dataset of Table 4.1

Person	Weight(x_n)	Height(y_n)	Class/Label	Euclidean distance
1	53	170	Underweight	16.03122
2	65	185	Healthy	16.49242
3	67	173	Healthy	**4.472136**
4	50	167	Underweight	19.10497
5	66	175	Healthy	**6.708204**
6	54	173	Underweight	15.52417
7	81	168	Obese	**12.04159**
8	85	170	Obese	16.03122

By applying Eq. 4.1, the Euclidean distance could be calculated as follows: Euclidean distance from person 1 is calculated as follows:

$$\sqrt{(169 - 170)^2 + (53 - 69)^2} = 6.08$$

In this way, Euclidean distance with respect to all data points is calculated, and the result is given in the following table (Table 4.2).

If k = 3, then the three nearest neighbors are 3, 5, and 7. Here, the majority of the neighbors are healthy. So the new person could be considered as a healthy one.

In recent years, k-NN has been extensively used in various engineering domain. In 2007, a novel methodology [11] was introduced by Tsang-Long Pao et al. to extract or recognize visual information from the audio-visual speech by weighted k-NN-based classification on Mandarin Database. They have done a comparative study based on three weighting functions with respect to weighted k-NN classifiers for recognizing digits from 0 to 9. They have studied three classifiers: basic k-NN, weighted k-NN, and weighted Dk-NN. Juraj Kacur et al. (2011) proposed a methodology [12] for identification of a speaker in a closed group. Dimension reduction methods like PCA and LDA were implemented prior to applying the k-NN classification.

They have demonstrated that by selecting a reduced number of eligible neighbors 6% improvement in speech recognition could be achieved. A novel methodology for speech emotion recognition for SROL database using weighted KNN algorithm was introduced by Feraru and Zbancioc [13]. They have used Wk-NN method to extract the vectors to the sentence level. Wk-NN is a slightly modified version of the traditional k-NN where each and every features vector is assigned a weight based on their performance in the classification process. A state-of-the-art methodology for phoneme prediction using k-NN and speaker ranking is proposed by Rizwan and Anderson [14]. k-NN was used to compute the speaker ranking by performing similarity analysis of the given test speaker with the instance space. They have done a comparative study of their experimental result with the nearest neighbor and k-NN majority voting approach. The proposed scheme has demonstrated improved prediction accuracy. This approach had an added advantage of on the fly customization of the speech recognizer for a given speaker and training data customization depending on the similarity measure.

4.2.2 Naïve Bayes (NB) Classifier

A typical disadvantage of using k-NN is, it cannot identify the important attributes since the classification is based on distance measure. Sometimes, it might happen that attributes which are less important having the same distance measure, considered as the important attributes. The major disadvantage of k-NN is, it cannot handle missing data. KNN is also a very slow procedure if it is acted upon very large data. Naïve based classifier applies Bayes theorem based on some independent set of features. It is suitable for high-dimensional input [15]. Naïve Bayes handles missing data effectively by excluding the attribute with the missing data at the time of calculating posterior probability. Naïve Bayes is an eager learning classifier, and it is much faster than k-NN. Conditional independence among the features is considered in Naïve Bayes and a maximum likelihood hypothesis is used. The notable characteristic of the Naïve Bayes model is that it learns over time.

Though Naïve Bayes is the most simple and basic classification method, it is a powerful one and can often outperform other sophisticated classification techniques. Extensive research on Naïve Bayes has been done since 1950s. In 1960s, Naïve Bayes became very popular for its text categorization ability [16].

The basic assumption of Naïve Bayes is that the features have an equal and independent effect on the outcome. Bayes theorem finds the probability of an event given the probability of another event which has been already happened. According to it, the probability of an event A given that B has been already happened could be obtained as follows:

$$P(A|B) = \frac{P(B|A)P(A)}{P(B)} \qquad (4.2)$$

where $P(A|B)$ = probability of an event A given that B is already happened, $P(A)$ is the prior probability of event A, and $P(B)$ is the probability of the event.

Let C be a set of mutually independent feature vectors and y is the decision or class variable.

Applying Bayes theorem, the following equation could be obtained:

$$P(y|C) = \frac{P(C|y)P(y)}{P(C)} \tag{4.3}$$

where $C = (c_1, c_2, c_3 \ldots c_n)$.

Naïve assumption is added with the Bayes which is simply the independence among the features. If A and B are the two independent events, then

$$P(A, B) = P(A)P(B) \tag{4.4}$$

Equation 4.4 could be applied to Eq. 4.2, and the following result could be obtained:

$$P(y|c_1, c_2, c_2, \ldots \ldots c_n) = \frac{P(c_1|y)P(c_2|y)\ldots\ldots P(c_n|y)P(y)}{P(c_1)P(c_2)\ldots\ldots P(c_n)} \tag{4.5}$$

Equation 4.5 could be written as

$$P(y|c_1, \ldots\ldots, c_n) = \frac{P(y)\prod_{i=1}^{n} P(c_i|y)}{P(c_1)P(c_2)\ldots P(c_n)} \tag{4.6}$$

Now since the denominator is constant for a particular input, it could be neglected

$$P(y|c_1, \ldots\ldots, c_n)\alpha P(y)\prod_{i=1}^{n} P(c_i|y) \tag{4.7}$$

A classifier model is built based on the Naïve Bayes theorem. The probability for a given set of inputs is computed, given the different values of the class variable y. The outcome having the maximum probability is selected for the final decision-making. The same could be represented mathematically as-

$$y = argmax_y P(y)\prod_{i=1}^{n} P(c_i|y) \tag{4.8}$$

Naïve Bayes classifier is illustrated by the following example.

Table 4.3 shows a sample training data which decides whether a person will play or not based on weather, temperature, and humidity of a certain day. Here, Naïve Bayes classification will be applied to determine whether a person will play or not

Table 4.3 Training data for playing condition

Weather	Temperature	Humidity	Play
Sunny	Hot	High	No
Sunny	Mild	High	No
Cloudy	Mild	High	No
Cloudy	Mild	Normal	Yes
Sunny	Cool	Normal	Yes
Rainy	Mild	High	No
Rainy	Cool	Normal	No
Sunny	Mild	Mild	Yes
Sunny	Mild	Low	Yes
Cloudy	Hot	High	No
Rainy	Mild	High	No
Cloudy	Cool	Normal	Yes

Table 4.4 (a) Conditional probability w.r.t Weather. (b) Conditional probability w.r.t. temperature. (c) Conditional probability w.r.t. Humidity

Weather	
$P(Sunny|Yes) = 3/5$	$P(Sunny|No) = 2/7$
$P(Cloudy|Yes) = 2/5$	$P(Cloudy|No) = 2/7$
$P(Rainy|Yes) = 0$	$P(Rainy|No) = 3/7$
Temperature	
$P(Hot|Yes) = 0$	$P(Hot|No) = 2/7$
$P(Mild|Yes) = 3/5$	$P(Mild|No) = 4/7$
$P(Cool|Yes) = 2/5$	$P(Cool|No) = 1/7$
Humidity	
$P(High|Yes) = 0$	$P(High|No) = 5/7$
$P(Normal|Yes) = 3/5$	$P(Normal|No) = 1/7$
$P(Mild|Yes) = 1/5$	$P(Mild|No) = 0$
$P(Low|Yes) = 1/7$	$P(Low|No) = 0$

on a certain day. Here, weather, temperature, and humidity are the feature variables and Play is the class variable or decision variable.

Let the test data is $X = \{$sunny, Hot, Normal$\}$, i.e., the system should determine whether a person will play if the weather is sunny, temperature is hot and humidity is low.

Table 4.4a, b and c shows the conditional probabilities w.r.t. class variable Play. The sample space $X = \{$Sunny, Hot, Normal$\}$.

By applying Eq. 4.7, the following relations could be obtained:

$$P(Yes|X) \approx P(Sunny|Yes).P(Hot|Yes).p(Normal|Yes).P(Yes)$$

$$\approx \frac{3}{5} \times 0 \times \frac{1}{7} \times \frac{5}{12} \approx 0$$

$$P(No|X) = P(Sunny|No).P(Hot|No).P(Normal|No).P(No)$$

$$= \frac{2}{7} \times \frac{2}{7} \times \frac{1}{7} \times \frac{7}{12} = 0.583$$

So one will not play if the weather is sunny, temperature is hot and humidity is normal.

In recent years Naïve Bayes classifier is applied extensively in music or speech recognition and classification. In 2010, Naïve Bayes classifier was explored by Zhouyu Fu et al. for music classification and retrieval [17]. From a local window, all audio features were extracted for classification instead of using a single level feature. They have studied two variants of Naïve Bayes classifiers based on the standard nearest neighbor and support vector machine classifiers. In 2012, Alberto Sanchis et al. proposed a Naïve Bayes classifier for confidence estimation in speech recognition [18]. Confidence estimation was used to classify a hypothesized word whether it was correct or not based on the extracted feature set. Here, a smoothed Naïve Bayes classification model was suggested to optimally combine the features. A combined feature extraction technique and Naïve Bayes classifier for speech recognition were proposed by Sonia Sunny et al. in 2013. They have developed a speech recognition system for identifying digits in Malayalam. Here wavelet methods like discrete wavelet transforms (DWT) and wavelet packet decomposition (WPD) were used for extracting the features. Once the features were extracted, the Naïve Bayes classifier was applied for the classification purpose. In 2016, a novel methodology for recognizing emotion based on the audio signal by applying Naïve Bayes classifier was proposed by Bhakre and Bang [19]. They have classified the audio signal into four basic emotional states by using Naïve Bayes classifier. Different statistical features such as pitch, energy, zero-crossing rate (ZCR), and mel-frequency cepstral coefficient (MFCC) were considered for the classification process.

4.2.3   Decision Tree and Speech Classification

The decision tree is a prediction-based analytical tool that has several applications in business analytics and critical decision-making processes. It is a kind of greedy algorithm that tries to identify an attribute to split a dataset based on certain preconditions. Decision trees are the examples of nonparametric supervised learning method that could be used for both classification and regression tasks. The decision rules are inferred from the data features, and thus a prediction model is built for finding the class label of the target variable. The decision tree is particularly useful for graphically representing the possible solutions depending on certain conditions that leads to a decision. It is a tree-like structure where the internal nodes are used to

Table 4.5 Training dataset showing mapping among weather attributes and playing condition

Climate	Temperature	Humidity	Windy	Play
Sunny	Hot	High	False	No
Sunny	Hot	Low	False	No
Sunny	Hot	Low	True	Yes
Cloudy	Cool	Low	False	Yes
Cloudy	Cool	Low	True	Yes
Cloudy	Cool	High	False	No
Rainy	Mild	Normal	False	Yes
Rainy	Mild	High	True	No
Rainy	Hot	High	True	No
Sunny	Mild	Normal	True	Yes

apply some test or conditions on an attribute. The branches represent the outcome of the test and leaf nodes hold the class label or final decision. This process is recursive in nature and applicable to all the sub-trees. If the class label is unknown for a certain tuple T then the same is predicted by the decision rules obtained from the decision tree.

There are several terminologies related to decision tree which are as follows.

Root Node: The entire sample dataset or population is represented by the root node. This could be further divided into homogenous subsets.

Leaf Node: The class labels reside in the leaf nodes. Leaf nodes could not be further segmented.

Splitting: The root node or non-leaf nodes could be further split into sub-nodes based on some conditions, generating different paths toward the class labels.

Branches or sub-trees: Branches or sub-trees are generated by the splitting procedure.

Pruning: Pruning is the process of removing the unwanted branches from the decision tree.

Iterative Dichotomiser 3(ID3) is a popular algorithm for constructing a decision tree invented by Quinlan [18]. ID3 is described in detail in the below section.

(i) **Iterative Dichotomiser 3(ID3):**

ID3 algorithm iteratively finds the splitting attribute among all the attributes in the sample space. Table 4.5 shows the training data space on which the ID3 algorithm would be applied.

Here based on four weather attributes, namely, climate, temperature, humidity, and windy, it will be decided whether a person will play or not.

In order to build a decision tree from the above training data, the first step is to choose the root attribute. Among all other attributes, the root attribute which is also called as the splitting node is selected based on the following two measures.

Entropy:
The randomness of the data is defined by the Entropy. It is a similarity measure that indicates the purity or impurity of the data. If the class label only indicates "Yes" or "No" then for a pure entropy the number of "Yes" will be equal to the number of "No's" or the class Label contains only "Yes" or "No". Entropy could be computed as given below [20, 21]

$$Entropy(S) = -P(yes)\log_2 P(Yes) - P(No)\log_2 P(No) \qquad (4.9)$$

If the number of yes is equal to the number of No, then the Entropy value reaches its peak. If the class label contains only pure data ("Yes" or "No"), then in both the case entropy value is zero. After calculating entropy, information gain is measured as follows:

$$Information\, Gain = Entropy(S) - [(Weighted\, Avg) * Entropy\, of\, Each\, Feature]$$
$$(4.10)$$

The attribute having the highest information gain is selected as the splitting attribute and the same process is recursively done in the subsequent attributes in each level. The overall ID3 algorithm is as follows.

ID3:
Algorithm

Step 1. Calculate the Entropy for the training dataset.

Step 2. For each attribute in the sample space or training dataset:

 (i) Calculate the entropy value considering all of the tuples.
 (ii) Take the average of the calculated entropy for the current attribute.
(iii) Compute the information gain.

Step 3. Choose the attribute having the highest information gain.
Step 4. Steps 1–4 are repeated until the desired tree is obtained.
Based on the above two measures the decision tree is built from the training dataset (Table 4.5) in the following way:
Total number of instances $= 10$.
Total number of "Yes" $= 5$.
Total number of "No" $= 5$.
From Eq. 4.9, the entropy could be calculated as follows:

$$Entropy\,(S) = -P(Yes)\log_2 P(Yes) - P(No)\log_2 P(No)$$

$$= -\frac{5}{14}\log_2\frac{5}{14} - \frac{5}{14}\log_2\frac{5}{14} = 0.32$$

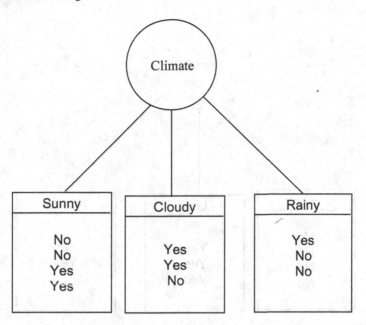

Fig. 4.1 Climate categories

The entropy of the entire dataset is calculated by the above procedure. Next step is to find the most suitable attribute for the root node or the splitting attribute. In order to do this, the information gain of each of the attribute is to be calculated.

We are continuing with the previous example. Under climate attribute, there are three categories which are shown in Fig. 4.1.

Entropy of each category is calculated by using Eq. 4.9.

$$Entropy\ (Sunny) = -\frac{2}{4} \log_2 \frac{2}{4} - \frac{2}{4} \log_2 \frac{2}{4} = 0.30$$

$$Entropy\ (Cloudy) = -\frac{2}{3} \log_2 \frac{2}{3} - \frac{1}{3} \log_2 \frac{1}{3} = 0.28$$

$$Entropy\ (Rainy) = -\frac{1}{3} \log_2 \frac{1}{3} - \frac{2}{3} \log_2 \frac{2}{3} = 0.28$$

From Eq. 4.10, information gain is computed as follows:

$$Information\ Gain = 0.32 - \left(\frac{4}{10} * 0.30 + \frac{3}{10} * 0.28 + \frac{3}{10} * 0.28 \right) = 0.44$$

In this way, the representation of the category Humidity and Windy is shown in Figs. 4.2 and 4.3 respectively. Corresponding Information Gain for Humidity and Windy are also calculated and shown below.

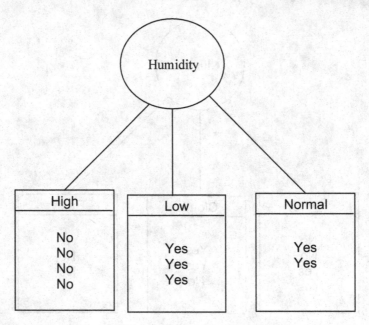

Fig. 4.2 Humidity categories

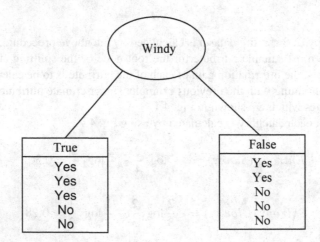

Fig. 4.3 Windy categories

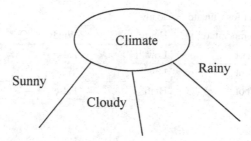

Fig. 4.4 Climate as root attribute

Table 4.6 Training data snippet for Climate = Sunny

Climate	Temperature	Humidity	Windy	Play
Sunny	Hot	High	False	No
Sunny	Hot	Low	False	No
Sunny	Hot	Low	True	Yes
Sunny	Mild	Normal	True	Yes

The attribute Humidity falls under pure Entropy since all the categories have either "Yes" or "No".

So $Entropy\,(Humidity) = 0$

The computation of entropy of Windy category is given below.

$$Entropy\,(False) = -\frac{3}{5}\log_2\frac{3}{5} - \frac{2}{5}\log_2\frac{2}{5} = 0.28$$

$$Entropy\,(True) = -\frac{3}{5}\log_2\frac{3}{5} - \frac{2}{5}\log_2\frac{2}{5} = 0.28$$

$$Information\;Gain = 0.32 - \left(\frac{5}{10} * 0.28 + \frac{5}{10} * 0.28\right) = 0.04$$

From the above calculations, it could be seen that the attribute climate has the highest information gain among all other attributes and hence selected as the first splitting attribute. This is shown in Fig. 4.4.

Under "sunny", the next splitting attribute should be selected from the remaining three, i.e., temperature, humidity, and windy. Table 4.6 shows the required sample space for sunny climate.

Again the total entropy of the system should be calculated from Eq. 4.9.

$$Entropy\,(S) = -P(Yes)\log_2 P(Yes) - P(No)\log_2 P(No)$$
$$= -2/4\log_2\frac{2}{4} - \frac{2}{4}\log_2\frac{2}{4} = -\log_2\frac{1}{2} = 0.30$$

Table 4.7 Data snippet for Climate = Cloudy

Climate	Temperature	Humidity	Windy	Play
Cloudy	Cool	Low	False	Yes
Cloudy	Cool	Low	True	Yes
Cloudy	Cool	High	False	No

For temperature:

$$Entropy\,(Hot) = -\frac{1}{3}\log_2\frac{1}{3} - \frac{2}{3}\log_2\frac{2}{3} = -0.15 - 0.12 = 0.27$$

$$Entropy(Mild) = 0$$

$$Information\,Gain = 0.30 - \frac{3}{4} * 0.27 = 0.1$$

For humidity:

$$Entropy\,(High) = 0$$

$$Entropy(Low) = 0.30$$

$$Entropy(Normal) = 0$$

$$Information\,Gain = 0.30 - \frac{2}{4} * 0.30 = 0.15$$

For windy:

$$Entropy\,(True) = 0 \; Entropy\,(False) = 0$$

$$Information\,Gain = 0.30 - 0 = 0.30$$

Information gain for the attribute Windy is greater than all other attributes; hence, this could be considered as the next splitting attribute under "Sunny". This is shown in Fig. 4.5.

For cloudy, the splitting attribute should be selected from the remaining attributes: "Temperature" and "Humidity". The data snippet is shown in Table 4.7.

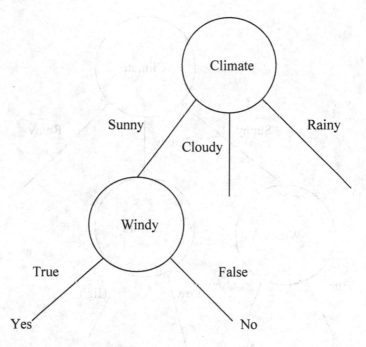

Fig. 4.5 The next splitting attribute "Windy"

Here total Entropy is

$$Entropy = -\frac{2}{3}\log_2\frac{2}{3} - \frac{1}{3}\log_2\frac{1}{3} = 0.27$$

For temperature:

$$Entropy\,(Cool) = -\frac{2}{3}\log_2\frac{2}{3} - \frac{1}{3}\log_2\frac{1}{3} = 0.27$$

$$Information\,Gain = 0$$

For windy:

$$Entropy\,(False) = -\frac{1}{3}\log_2\frac{1}{3} - \frac{1}{3}\log_2\frac{1}{3} = 0.31$$

$$Entropy\,(True) = 0$$

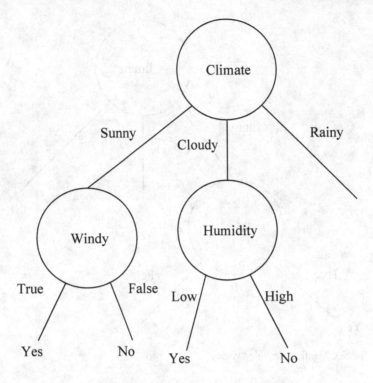

Fig. 4.6 Decision tree after adding humidity

$$Information\ Gain = 0.27 - \left(\frac{2}{3} * 0.31\right) = 0.06$$

For humidity:

$$Entropy\ (Low) = 0\ Entropy\ (High) = 0$$

$$Information\ Gain = 0.27$$

Since information gain of "humidity" is greater than "temperature", hence the next important splitting attribute is humidity. The decision tree after adding humidity is shown in Fig. 4.6.

The data snippet for climate "Rainy" is shown in Table 4.8

$$Total\ Entropy = -\frac{1}{3}\log_2\frac{1}{3} - \frac{2}{3}\log_2\frac{2}{3} = 0.28$$

Table 4.8 Data snippet for Climate = Rainy

Climate	Temperature	Humidity	Windy	Play
Rainy	Mild	Normal	False	Yes
Rainy	Mild	High	True	No
Rainy	Hot	High	True	No

For temperature:

$$Entropy\ (Mild) = -\frac{1}{3}\log_2\frac{1}{3} - \frac{1}{3}\log_2\frac{1}{3} = 0.31$$

$$Entropy\ (Hot) = 0$$

$$Information\ Gain = 0.28 - \frac{2}{3} * 0.31 = 0.07$$

For humidity:

$$Entropy(Normal) = 0 Entropy(High) = 0$$

$$Information Gain = 0.28$$

For windy:

$$Entropy(False) = 0 Entropy(True) = 0$$

$$Information\ Gain = 0.28$$

Here is a tie between Humidity and Windy. Both humidity and Windy is the viable candidate for the next deciding attribute under climate "Rainy". "Windy" is selected as the decision attribute under Rainy. The corresponding final decision tree is shown in Fig. 4.7.

From the decision tree, decision rules are constructed. Here, the attribute temperature is not a deciding attribute, so it does not appear in the decision tree.

One of the important properties of the decision trees is that there are no restrictions on the number or type of the features [22]. Multidimensional or multivariant data features could be effectively classified using a decision tree. Over the decades, the decision tree is used in various classification domains; speech recognition and classification is also not an exception. In 2012, observation likelihoods were computed for a predetermined hidden Markov model state by Akamine and Ajmera [23].

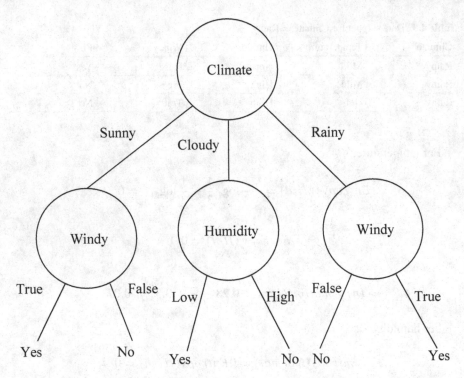

Fig. 4.7 Final decision tree

Instead of using Gaussian mixture models, decision tree was used to build the proposed model. The decision tree was applied for large vocabulary speech recognition. Dominic Telaarand and Fuhs.M. C. explored decision tree model to improve ASR system performance on South-Asian accented speakers in 2013 [24]. A new similarity metrics was proposed to analyze the different coarticulation across the accents and people and manifestation of that in the decision tree. One of the vital components of the automatic speech recognition (ASR) system is the acoustic model. Acoustic model defines the relationship between the acoustic observations and word sequences. In recent years, context-dependent deep neural network hidden Markov model (CD-DNN-HMM) has become the state-of-the-art acoustic modeling method for ASR [25–30]. In 2014, a decision tree-based state tying procedure was proposed by Li and Wu [31]. Here, sum-of-squared error was minimized by using the state embedding derived from DNN. Recently, music classification has become an important research interest due to its wide applicability. Use of decision tree for predicting musical genres classes are effective for monitoring ramification because nodes and branches of the tree are accessible. In 2017, Glaucia M. Bressan et al. proposed a decision tree based classification model for Latin musical genres [32].

4.2.4 Support Vector Machine (SVM) and Speech Classification

A support vector machine is a supervised machine learning algorithm that could be used for classification. It is a very powerful algorithm which has a wide range of applications ranging from text analysis to pattern recognition. The basic principle of SVM [33–35] is fixing a boundary to collect data points which are all alike or belongs to the same class. A new data point is compared with the boundary to know whether it is inside the boundary or not. The SVM method is generic in the sense that once a boundary is fixed then most of the training data becomes redundant. The data points which help to establish the boundary is called support vectors. Since these points support the boundary, these are referred to as support vectors. Each data point or observation is basically a data tuple containing values of different attribute or features, and hence, these are called "vectors". This boundary is traditionally called a hyperplane. To classify data points having two features, the hyperplane is basically a line separating the classes in two-dimensional spaces (shown in Fig. 5.8). The further the data points lie from the hyperplane, the more the probability that they are correctly classified. Whenever a new point has to be classified, it is checked on which side of the hyperplane it lands and that is the class of it.

Selection of the hyperplane is a crucial decision for any classification problem. There might be many hyperplanes between the data points, but an inappropriate selection of the hyperplane results in wrong classification. The thumb rule to choose a hyperplane is to select the hyperplane which has the greatest possible margin with the support vectors [36].

Real-world data is not as clean as shown in Fig. 4.8. In most of the cases, the data is jumbled up and thus becomes linearly inseparable (shown in Fig. 4.9a). Here, the classification problem becomes a tricky one. The concept is explained below assuming that the data points are black and white colored balls which are jumbled as shown in Fig. 4.9a. Figure 4.9b shows that the balls have lied on a two-dimensional plane. The problem is to divide the balls into two classes: white balls and black balls. Since all the balls are jumbled up it is not possible to separate them by a single line. Now if the black balls are tossed up, then a plane can separate them in three-dimensional spaces (shown in Fig. 4.9c). Now the hyperplane is not a single line rather it is a plane in three-dimensional spaces. Therefore, a two-dimensional problem is transformed into a three-dimensional one. The lifting of the balls is equivalent to the mapping of data to higher dimensions, which is also called kernelling.

More information about the theory of support vector machine could be obtained from "Support Vector Machines Succinctly" by Kowalczyk [37].

Support vector machines (SVMs) are the state-of-the-art classifiers. Since SVM is based on maximum margin solution it gives better result compared to most of the nonlinear classifiers in the presence of noise. Since ASR system is susceptible to noise, in recent years there is a growing interest in application of SVM in speech recognition and classification domain [38–42]. In 2011, a novel methodology for automatic speech emotion recognition was proposed by Shen et al. [43]. They have

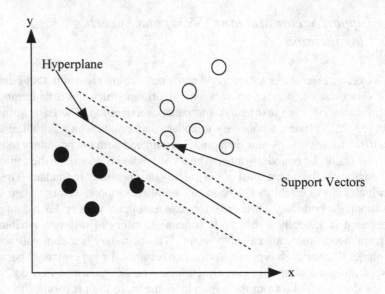

Fig. 4.8 Hyperplane in two-dimensional space

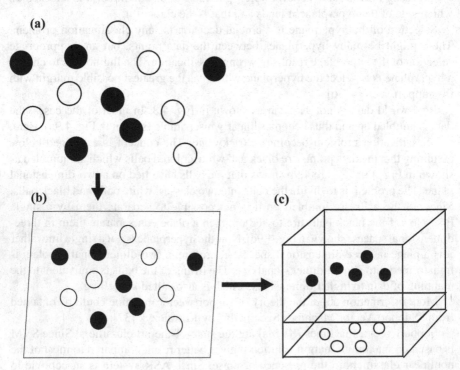

Fig. 4.9 **a** The jumbled up white and black balls. **b** The balls in two-dimensional space. **c** Hyperplane dividing the black and white balls in three-dimensional space

classified speech emotion into five classes such as boredom, disgust, sadness, happiness, and neutral. At first, features like energy, pitch, linear prediction coefficients (LPC), and mel-frequency cepstrum coefficients (MFCC) were extracted. SVM was applied to these features to classify different emotional states. In the same year, Davood Mahmoodi et al. suggested a methodology [44] for estimating age based on human's speech features. They have proposed an automatic age estimation system to classify Persian speakers belonging to six age groups. Perceptual linear predictive (PLP) and mel-frequency cepstral coefficients (MFCC) were used as features and SVM was utilized to do the classification. In 2013, a methodology for detecting incorrectly annotated words was proposed by Matoušek and Tihelka [45]. They have utilized SVM to detect wrongly annotated words. In 2016, Kamil Aida-zade et al. applied support vector machines to acoustic model of ASR system for Azerbaijani DataSet based on MFCC and LPC features [46]. They have analyzed various results generated during the training phase for different kernel functions. A speech classification method based on cuckoo algorithm and support vector machine was proposed by Wenlei Shi and Fan, X. in 2017 [47]. With the help of cuckoo algorithm, they have developed an optimized version of SVM(CS-SVM). Four types of music were selected, e.g., folk songs, Guzheng, pop, and rock. Speech features were extracted by using cepstral coefficients. CS-SVM was applied to train the characteristics signals, and thus, an optimal classifier model was developed.

4.3 Neural Network in Speech Classification

In Chap. 2, Sect. 2.3.2, a detailed discussion on hidden Markov model (HMM) and its role in speech recognition is given. Though HMM model has been a popular statistical model in speech recognition and classification for decades, from 1990s artificial neural network (ANN) has emerged as an effective alternative. A detailed discussion of neural network and its application in speech recognition has been done in Chap. 2, Sect. 2.3.3. This section discusses some of the pioneering works done on speech classification by applying artificial neural network over the times.

A model for classification of different speech accents was proposed by Chan et al. [48]. Several ANN models to recognize and identify accents were proposed here. A feasible approach was proposed for assisting ASR system to identify accents in an environment where multiple English accents exist. They have applied mainly three models: competitive learning [49], backpropagation [50], and counter propagation [51]. Twenty-two speakers were classified into two categories and three neural networks were implemented for the same task. Seven features were used as input to the neural network. The features used were the first three average formant frequencies, the average pitch frequency, percentage of duration of speaking English, the age at which English was first introduced and the number of years of speaking English. Among the three types of neural networks, the backpropagation ANN exhibited the superior performance. They have demonstrated the use of ANN to recognize native and non-native English speakers. In 1996, a neural network-based model of classification of speech under stress was proposed by Hansen and Womack [52]. The variation of vocal tract under stress is considered using cross-sectional areas, acoustic tubes, and spectral features. Based on the knowledge of these variations, a neural network-based classification model was formulated for 11 types of stress conditions: angry, clear, cond50, cond70, Lombard, fast, normal, loud, question, slow, and soft. A neural network trained with mono partition features (i.e., a single phone class partition) was used as the classifier. The individual speech feature partition was propagated through two intermediate hidden layers of the neural network. The stress probability score was estimated in the output layer. The existing speech under stress (SUSAs) database was used to evaluate the classification algorithm for selection of the feature set and stress classification algorithm performance. In 2001, a neural network based isolated word speech recognition system was proposed by Polur et al. [53]. They have chosen a small size vocabulary consisting of only words YES or NO. Cepstral analysis was done to extract spectral features from individual frames, and the extracted features were fed into a feedforward neural network. The network was trained to categorize the incoming words into respective classes. A pattern search function was used which accepts the output from the neural network. The pattern search function matches the input sequence with a given set of target word patterns. The purpose of this research was to build a humane machine interface that could be used to classify and identify the spoken utterances in normal condition or in impaired or distorted speech. In 2003, C. Shao and Bouchard M have done a comparative study of error rate of classification of noisy speech [54] by traditional classification algorithms with the neural network classifiers. They have used traditional classification algorithms like linear classifiers, nearest neighbor classifiers, and quadratic Gaussian classifier. Extended Kalman filter and the Levenberg–Marquardt algorithm were used to train the neural networks. They have demonstrated that more accurate and robust classification of noisy speech could be done by neural network classifiers rather than the traditional classifiers. A tailored neural network model to classify speech and nonspeech in hearing aids was developed by Enrique Alexandre et al. in 2008 [55]. They have explored the feasibility of using such a model for hearing aids. Some sorts of trade-off have to be considered keeping in mind about the two main important constraints regarding the hearing aids in market: computational complexity and battery life. Tailoring neural network was done by maintaining a balance which requires the reduction of computational complexity without hampering the performance of the classifier.

In the last decade, a growing research interest is observed on application of deep neural network on speech recognition. Some of the related pioneering works are discussed in the next section.

4.4 Deep Neural Network in Speech Recognition and Classification

A neural network having more than two layers is known as a deep neural network.

As discussed above, neural networks have a long history in speech recognition. In recent years mostly in the current decade, these have gained a significant attention with the advent of deep feedforward networks [56]. Because of the inherent dynamic characteristics of the speech recurring neural network or deep neural network could be considered as a suitable alternative in ASR system. A detailed discussion about deep neural network (DNN [57–61]) is given in Chap. 2, Sect. 2.3. In this section, some recent research trends about applying DNN is discussed.

A new methodology for acoustic modeling by deep belief network was proposed by Mohamed et al. in 2012 [62]. They have replaced Gaussian mixture models by deep neural networks and demonstrated that a better phonetic recognition could be achieved since DNN contains several layers of features and a huge number of parameters. Pretraining of these networks are done as a multiple layer generative model of a window of spectral feature vectors without making use of any discriminative information. After designing the features, slight adjustment of features was done by using backpropagation; therefore, better prediction of probability distribution over the states of monophonic HMM could be achieved. In 2013, application of deep neural network for both speaker and language recognition was proposed by Richardson et al. [63]. The primary contribution of this research was analyzing the application of a single DNN trained for ASR with respect to both speaker and language recognition. In 2015, a case study was done on music genre classification using deep learning methods by Rajanna et al. [64]. They have explored a two-layer neural network for music genre classification. They have done a comparative study on the accuracy rate of the DNN with the well-known traditional classification techniques, e.g., SVM, regression, etc. An improvised speech recognition system for stressed speech using deep neural network was proposed by Dumpala and Kopparapu [65] in 2017. The effect of including breath sound in the training phase of the speaker recognition was analyzed and compared by using the state-of-the-art Gaussian mixture model-universal background model (GMM-UBM) and DNN based systems. It was observed that the learning capability of the DNN based system is superior to the GMM-UBM based system even there is any unseen data.

4.5 Summary

This chapter introduces some state-of-the-art classification strategies for speech recognition and classification. The basic and most simple classification technique is k-nearest neighbor (k-NN) algorithm. It is useful when no clear idea about the distribution of data is available. This methodology was developed to perform discriminative analysis of data when no parametric estimation of probability densities is available or the same is difficult to estimate. After giving a vivid idea about k-NN, Naïve Bayes classification model is discussed. Speech recognition is a multi-class problem and Naïve Bayes classifier is effective to handle that. Naïve Bayes classifier is built on the Bayesian theory which is basically a simple and effective supervised classification technique based on probability estimation. The decision tree is a prediction-based analytical tool that has several applications in business analytics and critical decision-making processes. It is a kind of greedy algorithm that tries to identify an attribute to split a dataset based on certain preconditions. Since a decision tree does not have any restrictions on the number or type of the features it is effective to classify speech because of its multidimensional and multivariant features. Support vector machine is a powerful supervised classification technique that could be useful for speech recognition. SVM establishes a boundary between different classes in such a way that the margin between the boundary and the nearest points of each class would be maximized. SVM is effective to classify data points in a noisy environment because it is based on maximization of margin principle. Since ASR is susceptible to noise; SVM could be effectively used in speech classification. Though HMM model has been a popular statistical model in speech recognition and classification for decades, from 1990s artificial neural network (ANN) has emerged as an effective alternative. Some pioneering researches related to neural network speech classification model is discussed in this chapter. Deep neural network could be used as an effective alternative for the ASR systems because of inherent dynamic nature of the speech. Several researches have been done in the last decade related to the application of DNN in speech, and some of them are highlighted in this chapter.

References

1. Retrieved September 26, 2018, from https://www.youtube.com/watch?v=4HKqjENq9OU.
2. Retrieved October 22, 2018, from http://www.scholarpedia.org/article/K-nearest_neighbor.
3. Retrieved October 22, 2018, from https://www.jstor.org/stable/1403796?seq=1#page_scan_tab_contents.
4. Cover, T., & Hart, P. (1967). Nearest neighbor pattern classification. *IEEE Transactions on Information Theory, 13*(1), 21–27.
5. Hellman, M. E. (1970). The nearest neighbor classification rule with a reject option. *IEEE Transactions on Systems Science and Cybernetics, 3,* 179–185.
6. Fukunaga, K., & Hostetler, L. (1975). k-nearest-neighbor bayes-risk estimation. *IEEE Transactions on Information Theory, 21*(3), 285–293.
7. Dudani, S. A. (1976). The distance-weighted k-nearest-neighbor rule. *IEEE Transactions on Systems Science and Cybernetics*, SMC-6:325–327.
8. Bailey, T., & Jain, A. (1978). A note on distance-weighted k-nearest neighbor rules. IEEE Transactions on Systems, Man, Cybernetics, *8,* 311–313.
9. Bermejo, S., & Cabestany, J. (2000). Adaptive soft k-nearest-neighbour classifiers. *Pattern Recognition, 33,* 1999–2005.
10. Jozwik, A. (1983). A learning scheme for a fuzzy k-nn rule. *Pattern Recognition Letters, 1,* 287–289.
11. Pao, T. L., Liao, W. Y., & Chen, Y. T. (2007). Audio-visual speech recognition with weighted KNN-based classification in mandarin database. In *2007 Third International Conference on Intelligent Information Hiding and Multimedia Signal Processing, IIHMSP 2007* (Vol. 1, pp. 39–42). IEEE.

12. Kacur, J., Vargic, R., & Mulinka, P. (2011). Speaker identification by K-nearest neighbors: Application of PCA and LDA prior to KNN. In *2011 18th International Conference on Systems, Signals and Image Processing (IWSSIP)* (pp. 1–4). IEEE.

13. Feraru, M., & Zbancioc, M. (2013). Speech emotion recognition for SROL database using weighted KNN algorithm. In *2013 International Conference on Electronics, Computers and Artificial Intelligence (ECAI)* (pp. 1–4). IEEE.

14. Rizwan, M., & Anderson, D. V. (2014). Using k-Nearest Neighbor and speaker ranking for phoneme prediction. In *2014 13th International Conference on Machine Learning and Applications (ICMLA)* (pp. 383–387). IEEE.

15. Retrieved October 08, 2018, from http://www.statsoft.com/textbook/naive-bayes-classifier.

16. Russell, S., & Norvig, P. (2003). *Artificial intelligence: A modern approach* (2nd ed.). Prentice Hall. ISBN 978-0137903955. [1995].

17. Fu, Z., Lu, G., Ting, K. M., & Zhang, D. (2010). Learning Naïve Bayes classifiers for music classification and retrieval. In *2010 20th International Conference on Pattern Recognition (ICPR)* (pp. 4589–4592). IEEE.

18. Sanchis, A., Juan, A., & Vidal, E. (2012). A word-based Naïve Bayes classifier for confidence estimation in speech recognition. *IEEE Transactions on Audio, Speech, and Language Processing, 20*(2), 565–574.

19. Bhakre, S. K., & Bang, A. (2016). Emotion recognition on the basis of audio signal using Naïve Bayes classifier. In *2016 International Conference on Advances in Computing, Communications and Informatics (ICACCI)* (pp. 2363–2367). IEEE.

20. Quinlan, J. R. (1986). Induction of decision trees. *Machine learning, 1*(1), 81–106.

21. Retrieved October 11, 2018, from https://www.youtube.com/watch?v=qDcl-FRnwSU.

22. Navada, A., Ansari, A. N., Patil, S., & Sonkamble, B. A. (2011). Overview of use of decision tree algorithms in machine learning. In *Control and system graduate research colloquium (icsgrc), 2011 IEEE* (pp. 37–42). IEEE.

23. Akamine, M., & Ajmera, J. (2012). Decision tree-based acoustic models for speech recognition. *EURASIP Journal on Audio, Speech, and Music Processing, 2012*(1), 10.

24. Telaar, D., & Fuhs, M. C. (2013). Accent-and speaker-specific polyphone decision trees for non-native speech recognition. In *INTERSPEECH* (pp. 3313–3316).

25. Hinton, G., et al. (2012). Deep neural networks for acoustic modeling in speech recognition: The shared views of four research groups. *IEEE Signal Processing Magazine, 29*(6), 82–97.

26. Mohamed, A. R., Dahl, G., & Hinton, G. (2009). Deep belief networks for phone recognition. In *Nips workshop on deep learning for speech recognition and related applications* (Vol. 1, No. 9, p. 39).

27. Mohamed, A. R., Dahl, G. E., & Hinton, G. (2012). Acoustic modeling using deep belief networks. *IEEE Transactions on Audio, Speech & Language Processing, 20*(1), 14–22.

28. Jaitly, N., Nguyen, P., Senior, A., & Vanhoucke, V. (2012). Application of pretrained deep neural networks to large vocabulary speech recognition. In *Thirteenth Annual Conference of the International Speech Communication Association*.

29. Seide, F., Li, G., & Yu, D. (2011). Conversational speech transcription using context-dependent deep neural networks. In *Twelfth Annual Conference of the International Speech Communication Association*.

30. Dahl, G. E., Yu, D., Deng, L., & Acero, A. (2012). Context-dependent pre-trained deep neural networks for large-vocabulary speech recognition. *IEEE Transactions on Audio, Speech and Language Processing, 20*(1), 30–42.

31. Li, X., & Wu, X. (2014). Decision tree based state tying for speech recognition using DNN derived embeddings. In *2014 9th International Symposium on Chinese Spoken Language Processing (ISCSLP)* (pp. 123–127). IEEE.

32. Bressan, G. M., de Azevedo, B. C., & ElisangelaAp, S. L. (2017). A decision tree approach for the musical genres classification. *Applied Mathematics, 11*(6), 1703–1713.

33. Wang, Y., Cao, L., Dey, N., Ashour, A. S., & Shi, F. (2017). Mice liver cirrhosis microscopic image analysis using gray level co-occurrence matrix and support vector machines. Frontiers in artificial intelligence and applications. In *Proceedings of ITITS* (pp. 509–515).

34. Zemmal, N., Azizi, N., Dey, N., & Sellami, M. (2016). Adaptive semi supervised support vector machine semi supervised learning with features cooperation for breast cancer classification. *Journal of Medical Imaging and Health Informatics, 6*(1), 53–62.
35. Wang, C., et al. (2018). Histogram of oriented gradient based plantar pressure image feature extraction and classification employing fuzzy support vector machine. *Journal of Medical Imaging and Health Informatics, 8*(4), 842–854.
36. Retrieved October 10, 2018, from https://www.kdnuggets.com/2016/07/support-vector-machines-simple-explanation.html.
37. Kowalczyk, A. (2017). Support vector machines succinctly.
38. Padrell-Sendra, J., Martin-Iglesias, D., & Diaz-de-Maria, F. (2006, September). Support vector machines for continuous speech recognition. In *2006 14th European Signal Processing Conference* (pp. 1–4). IEEE.
39. Dey, N., & Ashour, A. S. (2018). Challenges and future perspectives in speech-sources direction of arrival estimation and localization. In *Direction of arrival estimation and localization of multi-speech sources* (pp. 49–52). Springer, Cham.
40. Dey, N., & Ashour, A. S. (2018). *Direction of arrival estimation and localization of multi-speech sources.* Springer International Publishing.
41. Dey, N., & Ashour, A. S. (2018). Applied examples and applications of localization and tracking problem of multiple speech sources. In *Direction of arrival estimation and localization of multi-speech sources* (pp. 35–48). Springer, Cham.
42. Dey, N., & Ashour, A. S. (2018). Microphone array principles. In *Direction of arrival estimation and localization of multi-speech sources* (pp. 5–22). Springer, Cham.
43. Shen, P., Changjun, Z., & Chen, X. (2011). Automatic speech emotion recognition using support vector machine. In *2011 International Conference on Electronic and Mechanical Engineering and Information Technology (EMEIT)* (Vol. 2, pp. 621–625). IEEE.
44. Mahmoodi, D., Marvi, H., Taghizadeh, M., Soleimani, A., Razzazi, F., & Mahmoodi, M. (2011). Age estimation based on speech features and support vector machine. In *2011 3rd Computer Science and Electronic Engineering Conference (CEEC),* (pp. 60–64). IEEE.
45. Matoušek, J., & Tihelka, D. (2013). SVM-based detection of misannotated words in read speech corpora. In *International Conference on Text, Speech and Dialogue* (pp. 457–464). Springer, Heidelberg.
46. Aida-zade, K., Xocayev, A., & Rustamov, S. (2016). Speech recognition using support vector machines. In *2016 IEEE 10th International Conference on Application of Information and Communication Technologies (AICT),* (pp. 1–4). IEEE.
47. Shi, W., & Fan, X. (2017). Speech classification based on cuckoo algorithm and support vector machines. In *2017 2nd IEEE International Conference on Computational Intelligence and Applications (ICCIA)* (pp. 98–102). IEEE.
48. Chan, M. V., Feng, X., Heinen, J. A., & Niederjohn, R. J. (1994). Classification of speech accents with neural networks. In *1994 IEEE International Conference on Neural Networks, IEEE World Congress on Computational Intelligence* (Vol. 7, pp. 4483–4486). IEEE.
49. Kohonen, T. (2012). *Self-organization and associative memory* (Vol. 8). Springer Science & Business Media, New York.
50. Rumelhart, D. E., & McClelland, J. L. (1986). Parallel distributed processing: explorations in the microstructure of cognition. volume 1. foundations.
51. Hecht-Nielsen, R. (1990). *Neurocomputing.* Boston: Addison-Wesley.
52. Hansen, J. H., & Womack, B. D. (1996). Feature analysis and neural network-based classification of speech under stress. *IEEE Transactions on Speech and Audio Processing, 4*(4), 307–313.
53. Polur, P. D., Zhou, R., Yang, J., Adnani, F., & Hobson, R. S. (2001). *Isolated speech recognition using artificial neural networks.* Virginia Commonwealth Univ Richmond School of Engineering.
54. Shao, C., & Bouchard, M. (2003). Efficient classification of noisy speech using neural networks. In *2003 Proceedings of Seventh International Symposium on Signal Processing and Its Applications* (Vol. 1, pp. 357–360). IEEE.

55. Alexandre, E., Cuadra, L., Rosa-Zurera, M., & López-Ferreras, F. (2008). Speech/non-speech classification in hearing aids driven by tailored neural networks. In *Speech, Audio, Image and Biomedical Signal Processing using Neural Networks* (pp. 145–167). Springer, Heidelberg.
56. Hinton, G., et al. (2012). Deep neural networks for acoustic modeling in speech recognition. *Signal Processing Magazine, 29*(6), 82–97. IEEE.
57. Wang, Y., et al. (2018). Classification of mice hepatic granuloma microscopic images based on a deep convolutional neural network. *Applied Soft Computing.*
58. Lan, K., Wang, D. T., Fong, S., Liu, L. S., Wong, K. K., & Dey, N. (2018). A survey of data mining and deep learning in bioinformatics. *Journal of Medical Systems, 42*(8), 139.
59. Hu, S., Liu, M., Fong, S., Song, W., Dey, N., & Wong, R. (2018). Forecasting China future MNP by deep learning. In *Behavior engineering and applications* (pp. 169–210). Springer, Cham.
60. Dey, N., Fong, S., Song, W., & Cho, K. (2017). Forecasting energy consumption from smart home sensor network by deep learning. In *International Conference on Smart Trends for Information Technology and Computer Communications* (pp. 255–265). Springer, Singapore.
61. Dey, N., Ashour, A. S., & Nguyen, G. N. Recent advancement in multimedia content using deep learning.
62. Mohamed, A., Dahl, G.E., & Hinton, G. (2012). Acoustic modeling using deep belief networks. *IEEE Transactions on Audio, Speech, and Language Processing, 20*(1), 14–22.
63. Richardson, F., Reynolds, D., & Dehak, N. (2015). Deep neural network approaches to speaker and language recognition. *IEEE Signal Processing Letters, 22*(10), 1671–1675.
64. Rajanna, A. R., Aryafar, K., Shokoufandeh, A., &Ptucha, R. (2015). Deep neural networks: A case study for music genre classification. In *2015 IEEE 14th International Conference on Machine Learning and Applications (ICMLA)* (pp. 655–660). IEEE.
65. Dumpala, S. H., & Kopparapu, S. K. (2017). Improved speaker recognition system for stressed speech using deep neural networks. In *2017 International Joint Conference on Neural Networks (IJCNN)* (pp. 1257–1264). IEEE.

Chapter 5
Conclusion

Audio/speech processing is a special case of digital signal processing (DSP), which is applied to process and analyze speech signals. Some of the typical applications of speech processing are speech recognition, speech coding, speaker authentication, speech enhancement, detection and removal of noise, speech synthesis, text to speech conversion, etc. This book provides a deep insight and in-depth discussion about audio processing and automatic speech recognition.

Audio is available from various sources like recordings of meetings, newscast, telephonic conversations, etc. In this era of information technology, with technological progress, more and more digital audio, video, and images are being captured and stored day by day. The amount of audio data is increasing exponentially on the web and other information storehouses. In order to efficiently use these huge multimedia data, there should be some effective search technique in use. In earlier days, it was difficult to interpret, recognize, and analyze the digitized audio using computers; therefore, organizations had to construct and manually analyze the written transcripts of the audio content. However, gradually with the advent of larger storage capacities, faster processing hardware, and better speech recognition algorithms, it is now possible to extract and index audio content efficiently using audio mining techniques. Audio indexing consists of finding good descriptors of audio documents which can be used as indexes to perform the archiving and searching. This book provides a detailed discussion about audio indexing and classical information retrieval problem. Two major audio indexing techniques, namely, text-based indexing or large vocabulary continuous speech recognition (LVCSR) and phoneme-based audio indexing are discussed here. In order to perform speech recognition, there is a need to understand human speech production system. This book provides a vivid description of human speech production mechanism and corresponding method of establishing an electrical speech production system. This will help in understanding the basics of artificial speech generation. Furthermore, the book elaborates the structure of a typical ASR system. A detailed discussion on acoustic analysis, acoustic models, language models, and pronunciation models is given here. Statistical models like hidden Markov model (HMM) and application of neural network in ASR systems are detailed out in this book. A central decoder is the brain of the ASR system to which output of all other models is fed and as a result, a huge number of search graphs consisting of individual HMMs are generated. One of the pioneering algorithms on this huge search space optimization is the Viterbi algorithm which is also discussed in detail in this book.

This book provides a detailed description of a number of common features of audio signals which are important for audio classification. Moreover, different audio feature extraction techniques like linear prediction coding (LPC), mel-frequency cepstral coefficient (MFCC), etc., are described in detail. One of the most important application and research areas of speech recognition is speech classification. This book summarizes some of the state-of-the-art classification techniques like "K-nearest neighbors", "Bayesian classifier", "decision tree", "support vector machine", etc. Neural network models have been used extensively in speech classification to increase the accuracy of classification to a great extent. This book also gives a brief idea about the speech classification and related neural network models for the further exploration of the researchers to apply the neural network models to improve the performance of different audio indexing parameters.

The primary focus of this book is to elaborate audio processing and speech recognition techniques and discuss the related research works. Currently, there are many cutting-edge researches going on the application of deep learning on automatic speech recognition. WaveNet technology is one of them which is basically an algorithm used to transform the raw input text data into audio. Using deep learning, there are some ongoing research works like lip reading and synchronization of lip movements with the input audio stream that have immense potential for improvement of the socioeconomic condition. Future work will include venturing into these types of application areas of machine learning algorithm and highlighting the relevant research work on them. As the new technologies of machine learning are continuously growing in the form of deep learning, bioinspired algorithms as well as other technologies like human–computer interaction, robotics, etc., there is always a scope to further enrich the audio processing methodologies by exploring these new techniques. The applications could be extended in the application areas like crime detection, music therapy, etc. As audio is the inevitable part of human life, different types of applications could be grown based on audio processing for the benefit of society.

Printed in the United States
By Bookmasters